COMPARATIVE BIOGEOGRAPHY

COMPARATIVE BIOGEOGRAPHY

DISCOVERING AND CLASSIFYING BIOGEOGRAPHICAL PATTERNS OF A DYNAMIC EARTH

Lynne R. Parenti
and
Malte C. Ebach

RAWAT PUBLICATIONS
Jaipur • New Delhi • Bangalore • Guwahati • Kolkata

The authors and publisher gratefully acknowledge the generous contributions to this book by the Smithsonian Institution.

ISBN 978-81-316-0825-8

Indian Reprint, 2017

Published by
Prem Rawat for **Rawat Publications**
Satyam Apts., Sector 3, Jawahar Nagar, Jaipur - 302 004 (India)
Phone: 0141 265 1748 / 7006 Fax: 0141 265 1748
E-mail : info@rawatbooks.com
Website: rawatbooks.com

New Delhi Office
4858/24, Ansari Road, Daryaganj, New Delhi 110 002
Phone: 011-23263290

Also at *Bangalore, Hyderabad, Guwahati* and *Kolkata*

Printed and bound in India.

In Memory of Donn Eric Rosen (1929–1986)

Contents

Preface

How do we discover patterns in nature and what kinds of information and methods do we use to describe, evaluate, and interpret those patterns and compare them to others, including those of Earth history? This is what this book is about. It is aimed at the student of biogeography, yet written for all natural historians. We are both biologists, one a zoologist, the other a paleontologist; and we are both biogeographers. To many, these are non-overlapping, or mutually exclusive, categories; one may be a zoologist or paleontologist first and a biogeographer second. This is not our experience; it is not the way we work. We are both biologists and biogeographers at the same time.

Biogeography is a critical tool for discovering and interpreting the patterns and processes of life. A paper proposing a new hypothesis on the relationships of a group of organisms suggests other questions to biogeographers: "Where do they live?" Maps and habitat descriptions are as important as scientific illustrations or data matrices. "What do they live with?" The distribution and relationships of the entire biota is paramount, not corollary. Although not geologists, biogeographers consider patterns of Earth history, the hypotheses, the type and the amount of data that support geological patterns. Biogeographers can help geologists because biological patterns elucidate Earth history as readily as geology can inform biology.

Biogeographers are also historians, and they study not just the history of the Earth and the life on it, but also the history of ideas on how we should interpret this information. Ultimately, biogeographers are "natural historians with maps"—the moniker that best describes who we are and what we do.

Biogeography is a classic field of investigation with a rich history. All naturalists have been biogeographers, whether they noted which bird perched on what rock or described the coherence of life around the Pacific basin. That deep tradition acknowledges that biogeographers of the past wrestled with many of the same issues we still puzzle over. This book is our view of biogeography—a tumultuous field with many methods and roiling controversies, many of which are reviewed here. A student new to biogeography may be astonished to learn that there is such discord in a field that is often defined simply as the study of the geographic distribution of plants and animals. "What lives where and why?" seems a direct question. It may be answered in many ways depending on the kinds of assumptions one makes about organisms and their relationships to their environment and their evolutionary history. It may also be answered in different ways depending on the assumptions one makes about the data used to generate an areagram—morphological versus molecular—and how an areagram (area cladogram) is interpreted. As we will describe, even how we define *areas* will affect our biogeographic interpretations.

Understanding the pattern of biotic divergence is the key to discovering and reconstructing former geographical, climatic, and ecological mechanisms or processes that were responsible for biotic evolution. This book is a comprehensive history of biogeography, its theories and methods, and developments that have led to formulation of the principal methods of what we call comparative biogeography. We begin our text in the late 18th century to set the stage for the upheavals in our understanding of the history of Earth and its biota that were soon to follow. With this background we ask, what triggers evolution of biodiversity and distribution patterns? Empirical examples question the way geologists look at the evolving Earth. Our goal is to demonstrate that comprehensive cladistic methods can extract patterns resulting from interaction with Earth processes at all scales, helping us to understand what events have caused speciation and extinction and where these events may have taken place. This unifying research theme in systematics and evolutionary biology is central to tackling the current and anticipated continued large influx of molecular data and attendant cladograms and for establishing a biogeographic framework for the next generation of research scientists.

ACKNOWLEDGMENTS

Discussions over the years with Gary Nelson, Chris Humphries, and David Williams helped shape many of our ideas on biogeography. Michael Heads and David Williams kindly read a complete draft of the book manuscript and provided valuable criticism and extensive comments, for which we are most grateful. Two anonymous reviewers posed pointed questions that we hope we have answered to their satisfaction. Other colleagues graciously read and commented on draft chapters: Roger Buick, Tony Gill, John R. Grehan, Daniel Rafael Miranda-Esquivel, and Juan Morrone. Victor G. Springer provided innumerable, valuable references on biology and geology.

Chuck Crumly, of UC Press, patiently guided us through all stages of this book and urged us to write clearly and economically. Francisco Reinking, also of UC Press, expertly oversaw book production. Production of the maps and color reproduction of the figures was supported by the office of Richard Vari and Don Wilson, successive Chairs of the Department of Vertebrate Zoology, National Museum of Natural History, Smithsonian Institution.

Jacques Ducasse, Nathanaël Cao, and René Zaragüeta, of the Université Pierre et Marie Curie-Paris 6, Laboratoire Informatique et Systématique, provided the systematics and biogeography software *NELSON05* and answered our many questions about its use. Their support was critical for the completion of this book.

Illustrations were generously supplied, or permission to reproduce those previously published was granted, by Cheryl Backhouse and the Centre for Plant Biodiversity Research, Australian National Botanic Gardens; Tim Berra; John Brill; Ricardo Calejas; Giesela Dohrmann; Simone Farrer and Carla Flores, of CSIRO Publishing; Adrian Fortino; Chris Glasby; Joy Gold; Sam Gon III; John R. Grehan; Mike Hadfield; Peter Hovenkamp; Caitlin Hulcup; Ben Kennedy and Oxford University Press; Carolyn King; Evan Mantzios; Paul Muir; Rod Page; Lisa Palmer; Theodore Pietsch; David Reid and the *Journal of Molluscan Studies* (formerly *Proceedings of the Malacological Society of London*); Beatriz Rivera; Erin Clements Rushing and the Smithsonian Institution Libraries; Rick Sardinha; Georgina Steytler; Wesley Thorsson; Ángel Viloria; Carden Wallace; Judy West; and Tony Windberg. Base maps were prepared by Dan Cole and expertly rendered by Colleen Lodge. Other technical assistance was provided by Jeffrey M. Clayton.

Tina Ramoy, and Ella, remind LRP every day that there's no place like home.

ACKNOWLEDGMENTS

ONE

INTRODUCTION

> Laws of distribution can only be arrived at by comparative study of the different groups of animals, for this study we require a common system of regions and a common nomenclature.
>
> **Alfred Russel Wallace (1894:612)**

CLASSIFICATION IN SCIENCE

Biogeography is a comparative science. Classification is the foundation of comparative science. Whenever we compare two objects, we rely on a classification to decide whether they should be placed in the same group or in different groups. A scientific classification has two qualities (Szostak, 2005:2): it should first identify an exhaustive set of types, such as the Periodic Table of chemical elements, and second be based on some

Box 1.1 Universal Systems and General Laws

- A *Universal System* is an inclusive plan, arrangement, or classification that is characterized by repeatability and predictability. The Periodic Table of the Elements is a Universal System.
- A *General Law* is an immutable expression of the relationship among a series of observations. The notion that gradual changes in the Earth over long periods of time explain the origin and history of biodiversity and geodiversity is a General Law.

theoretical ordering principle, such as atomic number. In physics, the classification of colors was pioneered in the symmetrical six-color circle or wheel of German poet, writer, and naturalist Johann Wolfgang von Goethe, first published in *Zur Farbenlehre* (1808–1810; Goethe, 2006). Goethe's color wheel is still used today in science and industry, often in a modified form, such as the circular chart of Munsell (1905; see Platts, 2006). The elegantly simple color wheel represents a scientific classification. It incorporates the range of colors in the visible spectrum and places them in order of wavelength: red, orange, yellow, green, blue, purple. In geology, classifications are essential for the identification of rocks and of their minerals. *A System of Mineralogy, Fourth Edition,* by 19th-century naturalist James Dwight Dana (1854) introduced a chemical classification, grouping minerals into now familiar categories such as sulfides, silicates, and oxides (Hawthorne, 1985; see Ferraiolo, 1982).

The value of such natural classifications is that they accommodate *all possible* histories; hence, they are *universal* and have great predictive (or retrodictive) value. Without natural classifications, we cannot make meaningful comparisons of biological, chemical, or physical forms. Individual histories are not universal and cannot be used to classify or compare forms. The swimming performance of a particular species of tuna, for example, cannot tell us whether other fish species swim as fast or as slowly. Such individual histories play almost no role in predicting what other histories may be discovered. Knowing what other species are classified in the tuna family allows predictions about form and function of those species.

To understand a vast and complex system of interactions, we generalize our observations and experiences to recognize either *Universal Systems* or *General Laws*. A General Law is resistant to other possible explanations and can reject a Universal System. Scientific classifications should be Universal Systems, not General Laws. Classifications that are Universal Systems provide a stable foundation for all scientific fields.

Box 1.2 Léon Croizat (1894–1982)

Figure 1.1. Léon Croizat in Coro, Venezuela, August 1977. [Photograph courtesy of Ricardo Callejas and Beatriz Rivera.]

Léon Croizat is the father of modern biogeography. He formalized the concept of a dynamic Earth evolving along with the organisms that inhabit it, now sometimes called panbiogeography. Croizat, an Italian emigrant to the United States, was employed as a technical assistant at the Arnold Arboretum, Harvard University, from 1937 to 1946. In 1947, he moved to Venezuela, where he held several academic positions and worked on various field expeditions as a botanist, including his first exploration of the upper Orinoco. His experience and skill as a field botanist and scholar led him to write several groundbreaking works: *Manual of Phytogeography* (1952), *Panbiogeography* (1958), *Principia Botanica* (1961), and *Space, Time, Form: The Biological Synthesis* (1964) (see Craw, 1984). *(continued)*

Croizat's panbiogeography was an advance in comparative biogeography as it focused on organisms and geographical areas as distinct, yet interactive, entities. *Panbiogeography* was unique as most other biogeographical fields were developed against a backdrop of a static or slowly changing Earth. Croizat's idea that organisms naturally disperse and become geographically isolated within existing geographic ranges gave birth to the concept of vicariance (see Chapter 5). Croizat is a controversial figure in biogeography; the importance of his contributions continues to be debated (e.g., Seberg, 1986; Craw et al., 1999; Grehan, 2006). His extensive writings are most appreciated by those who take the time to read them: Croizat's ". . . flood of words has raised the sea of biogeography to a new level. . . . [His] victory is the defeat of hypotheses of chance dispersal: he has given us whereon to stand" (Corner, 1963:244–245).

EARTH AND LIFE EVOLVED TOGETHER

The catchphrase of Léon Croizat (1964:605), ". . . earth and life evolve together," refers to the dynamic interaction of biology and geology—a cornerstone of panbiogeography and one of the principles that we and many other biogeographers have adopted.[1] The concept of an Earth that changes along with the organisms that inhabit it has been controversial and is far from universally accepted as part of the foundation of biogeography. British geologist Charles Lyell outlined a General Law on the history of the Earth in his three-volume masterpiece, *Principles of Geology* (1830–1833). Lyell's General Law of gradual change over long periods of time was used to explain how the Earth was formed and to explain the origin and history of *biodiversity* and *geodiversity*. We call it a General Law because it was resistant to and rejected other possible explanations. Noted for his profound influence on geology, Lyell was one of the first to propose an explicit dispersalist biogeography which maintained that evolution of the Earth and distribution of life on it are disjunct mechanisms (see Camerini, 1993:705; Bueno-Hernández and Llorente-Bousquets, 2006). As we shall explain, the influence of such strict dispersalist views impeded progress in the science of biogeography.

The proposal of continental drift by German scientist Alfred Lothar Wegener (1915, 1929) diminished Lyell's notion of gradualism. Continental drift is a theory of Earth history based on the outline and positional relationships of continents as evidenced by the relationships of their biological and geological diversity. A supercontinent, Pangea, was formed and then subsequently broke apart, and over millions of years its sections or continents drifted to the positions they occupy today.[2] Late-19th- and early-20th-century biologists were intrigued by the growing evidence for past continental connections and interpreted the biogeographic patterns with respect to Earth history: Irish naturalist Robert Scharff's (1911) monumental *Distribution and Origin of Life in America* is a modern-in-tone refutation of the permanence of ocean basins and an argument for past land connections. Wegener's theory of continental drift was rejected by early-20th-century geologists, and hence by most other scientists as well, because he proposed no plausible explanatory mechanisms of continental formation or movement. The discovery of spreading mid-oceanic ridges in the mid-20th century vindicated Wegener and led to the proposal of a mechanism of an evolving Earth: plate tectonics and seafloor spreading (Hess, 1962). A new geological synthesis, incorporating a dynamic Earth, was adopted rapidly by geologists and other scientists (e.g., Dietz and Holden, 1970; Hallam, 1973). The development of a theory of plate tectonics dramatically altered our understanding of the Earth and changed perspectives on the patterns and mechanisms of extinction and evolution of life (see also Heads, 2005a). A dynamic Earth—not the passive, slowly eroding Earth, punctuated by catastrophes, as perceived by Lyell and other 19th-century naturalists—is taken for granted today.

Ironically, biogeography was led by 19th-century naturalists who gave in to the concept of a static Earth after considering a mobilist perspective (see also Chapter 2). British naturalist and biogeographer Alfred Russel Wallace, co-proposer with Charles Darwin of a theory of organic evolution (i.e., natural selection), argued first that geographical relationships of plants and animals, as detailed on maps, ". . . provided the crucial link between biological processes (the production of new species from existing ones) and geological processes . . ." (Camerini, 1993:723). Wallace even advocated major continental movements, but then changed his mind, as explained by Camerini (1993:726), who notes that Wallace argued,

> Just as geological and physical features provide clues to biological evolution, the evolutionary relationships and geographical distribution of animals provide essential clues to former land connections. On this point, however, we find in 1863 a shift from the reliance on major continental movements to a belief in the permanence of the major continental land

> masses. . . . The pro-permanence view provided solid ground for [Wallace's] subsequent treatises on geographical distribution and earned him the full support of Lyell and Darwin.

Had Wallace maintained a mobilist view of the world and its biota, we could now be in the second century of discovery of biogeographic patterns that incorporate Earth history. Instead, following in the Wallace-Darwin-Lyell biogeographic tradition, overwhelming biogeographical patterns that link continents, such as coherence of life around the Pacific basin, have been explained away as being irrelevant or as being driven by mechanisms such as long-distance dispersal of individual clades (e.g., Darlington, 1957; Carlquist, 1965; Briggs, 1974; and more recently de Queiroz, 2005; see Chapter 7). Geology and biology have been kept apart.[3]

Explanatory Mechanisms

Development of the phylogenetic systematic or cladistic methods (Hennig, 1950, 1965, 1966; Nelson and Platnick, 1981) to discover and rigorously diagnose monophyletic groups of organisms—and hence to build natural biological classifications—has been the greatest advance in evolutionary biology since the modern synthesis combined genetics with biological evolution (Mayr, 1942), and in systematic morphology since the reestablishment of Owen's special homology by Naef (1919). Biological classification changed in the mid-20th century in response to the rise of cladistic methodology rather than in response to the modern synthesis (e.g., Mayr, 1974; Ragan, 1998) or to the acceptance of the notion of a dynamic Earth. Biological classification, once largely gradistic, was replaced by a phylogenetic or cladistic classification system. In a cladistic classification, only monophyletic groups are named; in a gradistic classification, paraphyletic as well as monophyletic groups are named. A paraphyletic group, such as the Algae, Invertebrata, or Reptilia, is an artificial and non-evolutionary category that cannot be used to explain phylogenetic history.

Today biogeographic theories acknowledge the decisive role of phylogeny. Multiple phylogenies are mandatory to identify patterns. Without a biogeographical classification that incorporates natural biotic area groups based on a phylogenetic classification, we must explain each incidence of conformation to a pattern *as if it were not part of the pattern*. In effect, we give up the opportunity to compare. One pattern

Box 1.3 Cladistic versus Gradistic Classification

- Cladistic Classification: A biological classification in which only monophyletic groups are named. A monophyletic group, or *clade,* contains all, and only, the descendants of a common ancestor.
- Gradistic Classification: A biological classification in which names may be applied to both monophyletic and paraphyletic groups, emphasizing the differences among taxa. A paraphyletic group, or *grade,* contains descendants of a common ancestor yet excludes those descendants that have diverged from their close relatives.

If taxon A evolves into taxon B, all members of taxon A are paraphyletic because some members of taxon A are more closely related to members of taxon B than they are to any other taxon. If we assume that ancestors are found at the nodes of phylogenetic trees, then groups at the terminal branches are *grades,* not *clades.*

expressed by many different organisms is meaningful and has predictive power, even without a ready explanatory mechanism. The problem with particular explanatory distributional mechanisms—such as individual episodes of long-distance dispersal—is not that they fail to explain distributions and biodiversity, but that these explanations cannot be refuted empirically. Such explanatory hypotheses lack empirical rigor and are untestable. Only with a natural classification of taxa and biotic areas are we able to compare distributions and discover historical biogeographical patterns. This is what this book is about.

Life and Earth and Earth and Life

Clades form phylogenetic patterns because they share a common history and, therefore, their homologous characters (such as feathers in birds, seven cervical vertebrae in mammals, spinnerets in spiders, and so on). Abiotic patterns in geology involve structure and composition of minerals and rocks; their explanations perhaps allude to similar conditions under which they were formed (such as sedimentation or volcanism) or to geomorphological structures (asymmetrical and symmetrical rippling). These inorganic classifications reflect the types of environments that existed, but the structures are not necessarily related by common history. Ripples like those we see in coastal inlets or in tidal rivers are similar to the ripples we see in sedimentary rocks. Discovery of such rocks in association with other, similar structures that indicate a coastal or tidal environment leads to identification of patterns suggesting, in turn, that a current mountainous or arid terrane may have once been a coastline. Thus, ripples in various places or at different times are caused by similar mechanisms, but may not be caused by precisely the same event.

Inorganic classifications, furthermore, can never provide evidence as to what taxa lived in past environments. If we identify a coastal environment based on geological or geographical evidence, it still cannot tell us what families of fishes or gastropods, for example, may have lived there. No matter how much we know of a past environment, even its chemical or climatic composition, those data alone will not confirm what taxa lived there. The fossil record has shown in many instances that similar environments can support many different types of biota through time.

BIOGEOGRAPHY

The word *biogeography* was coined by the German geographer Friedrich Ratzel (1891:9):

> . . . Vereinigung der Pflanzen- und Tiergeographie mit der Anthropogeographie zu einer allgemeinen Biogeographie, einer Lehre von der Verbreitung des Lebens auf der Erde.
>
> . . . the unification of plant and animal geography with anthropogeography [human geography] in order to form a General Biogeography, the study of the distribution of life on Earth.

Ratzel's General Biogeography possibly combines all known methods, theories, and techniques of biogeography, including human geography, anthropology, and social change (see Müller, 1995). Here, we limit biogeography to the study of the relationship between the organic part of the world, the biosphere, and the inorganic, the physical Earth. The timeframe of biogeography spans nearly 4 billion years, from when life first appeared on Earth as simple cells, to the present day. In practice, biogeography does not extend much beyond some 570 million years ago (mya), when organisms became more complex and evolved hard parts that could be fossilized (see Tarling and Tarling, 1975).

Biogeography is a naturally integrative field of study that encompasses a broad range of methods, data, habitats, and organisms, as well as practitioners and goals.[4] Biogeography helps us understand our planet and its geography, geology, and organisms, where they have interacted through time, evolving together to form the places we know today. Most important, biogeography is a comparative science that interprets the complexity of relationships and distributions of life on Earth with respect to its geological history.

The common goal of all biogeographers is to understand the relationship between life and its distribution. After that, agreement is infrequent (see, e.g., Crisci, 2006; Crisci et al., 2003, 2006). Cladistic methods have been applied to biogeography in a variety of methods, many with contradictory aims (viz. Nelson and Platnick, 1981; Morrone and Carpenter, 1994; Humphries and Parenti, 1999; Brooks and McLennan, 2002; see Chapters 5 and 6). The method of Comparative Biogeography and its incorporation of Systematic and Evolutionary Biogeography is introduced below and detailed more fully in Chapters 3, 4, and 7. Other biogeographic methods and distributional mechanisms are reviewed with respect to the comparative biogeographic method in Chapters 5 and 6.

COMPARATIVE BIOGEOGRAPHY

Biogeography can be a powerful tool to explore data on the diversity, phylogeny, and distribution of organisms, to reveal the biological and geographical history of Earth. We aim to unite the many aspects of biogeography under one banner: *Comparative Biogeography*. Comparative biogeography uses the naturally hierarchical phylogenetic relationships of clades to discover the biotic area relationships among local and global biogeographic regions. One biotic area, A, may be said to be related to another, B, more closely than either is to a third, C, if the taxa of the biotic areas reflect a three-area relationship: C(AB). Such proposals of area relationship are three-area relationships or *area homologs:* hypotheses of area relationships that may be expressed in a general classification of areas.

To introduce comparative biogeography, we differentiate between the two types of biogeographic investigation that it encompasses: systematic biogeography and evolutionary biogeography.

Systematic Biogeography is the study of biotic area relationships and their classification and distribution. For example, the distribution and relationships of numerous taxa may be expressed in a hierarchy as Eastern South America (Africa, India), meaning that organisms in Africa have their closest relatives in India and that together they are in turn related to organisms in eastern South America. Examples include such diverse taxa as vascular plants, fishes, birds, and dinosaurs.

Evolutionary Biogeography is the proposal of evolutionary mechanisms responsible for organismal distributions. Possible mechanisms responsible for the distribution of organisms related as in the area homolog Eastern South America (Africa, India) include widespread taxa disrupted by continental break-up or individual episodes of long-distance movement, to name just two.

Systematic versus evolutionary is one historical division of biogeographers as well as of biologists and their methods. The division is analogous to investigation of "classification versus explanation" or "patterns versus mechanisms." This division dates from the earliest formulations of evolutionary theory (Camerini, 1993; see Chapter 2). The modern synthesis emphasized process or mechanism over pattern, and, according to Ghiselin (2006), at the level of species or below, with scant concern for geological processes (Chapter 2). Evolutionary biology under the modern synthesis did not focus largely on a dynamic Earth, emphasizing instead mechanisms such as dispersal, and species interactions, such as competition, mutualism, and predation. The dynamic Earth is more than just drifting and colliding continents; it is all the geological processes linked explicitly to events such as climate change, sea level changes, erosion and weathering, frequent volcanism, earthquakes, tidal waves, changes in atmospheric chemistry, changes in soil chemistry, and so on. Ultimately, it involves the close relationship between organisms and the environment, seen in major animal constructions such as coral reefs, and acting at all levels.

CLASSIFICATION OF AREAS: SYSTEMATICS AND BIOGEOGRAPHY

Classification of biotic areas is the goal of our comparative biogeography just as classification of taxa is the goal of systematics (Chapter 2). Biotic areas are what we compare and classify in biogeography. Biotic areas are defined by both the aggregate taxa and the areas in which they live.

Once comparative biogeography is more fully implemented, we will be able to replace the traditional classifications of biogeographic regions and realms with natural, homologous areas (sensu Wallace, 1894; Chapter 2). Arbitrary areas (e.g., an abiotic geographic entity, such as "Australia," "Borneo," or "the Philippines") have little meaning in comparative biogeography unless they are occupied by a monophyletic biota. For a variety of reasons, it is not surprising that some geographically delineated regions may also be recognized in area homologs.

TOWARD A COMPARATIVE BIOGEOGRAPHY

Comparative biogeography provides biologists with a rigorous empirical theory and with methods of analysis for interpreting Earth history. Comparative biogeography diagnoses and classifies biogeographic areas by incorporating data from a broad array of taxa, their phylogenetic hypotheses, and geological and geographic variables. It grapples with the potentially enormous amounts of data of comprehensive biogeographical analyses by providing a classification of organic areas which forms a biogeographical framework. It implements a classification or universal system *before* exploring explanatory mechanisms or hypotheses.

Comparative biogeography empirically examines the common historical processes that may be postulated to explain biotic distributions and diversity. It does not emphasize molecules over morphology, nor does it emphasize vicariance over dispersal (Chapter 5). This search for common patterns does not emphasize the simple over the complex (viz. Brooks, 2005). Simple mechanisms can produce highly complex, repetitive, nested patterns (Wolfram, 2002). Nature endlessly repeats (Croizat, 1958). This repetition, the observation of the same distribution over and over, in many different and unrelated taxa, led to the identification of natural biogeographic features which Croizat illustrated as lines on maps or tracks (Figure 1.2; also Chapter 2). Tracks drawn as networks or reticulations do not identify area homologs. The repeated features of global biogeography, trans-Pacific, trans-Atlantic, boreal, austral, Indian Ocean, and so on down to the lowest levels, when defined as area homologs and classified hierarchically, will form the framework of a comparative biogeography.

Biogeographers have swung between two extremes, from rejecting geological history as too old to have affected biological distribution, to interpreting distributions explicitly with respect to current theories of geology. We adopt the view of early cladistic biogeographers such as American ichthyologist and biogeographer Donn Eric Rosen (1978; Chapter 7), who states that biological and geological patterns provide "reciprocal illumination" or shed light on each other, but do not test, and therefore cannot reject, one another.

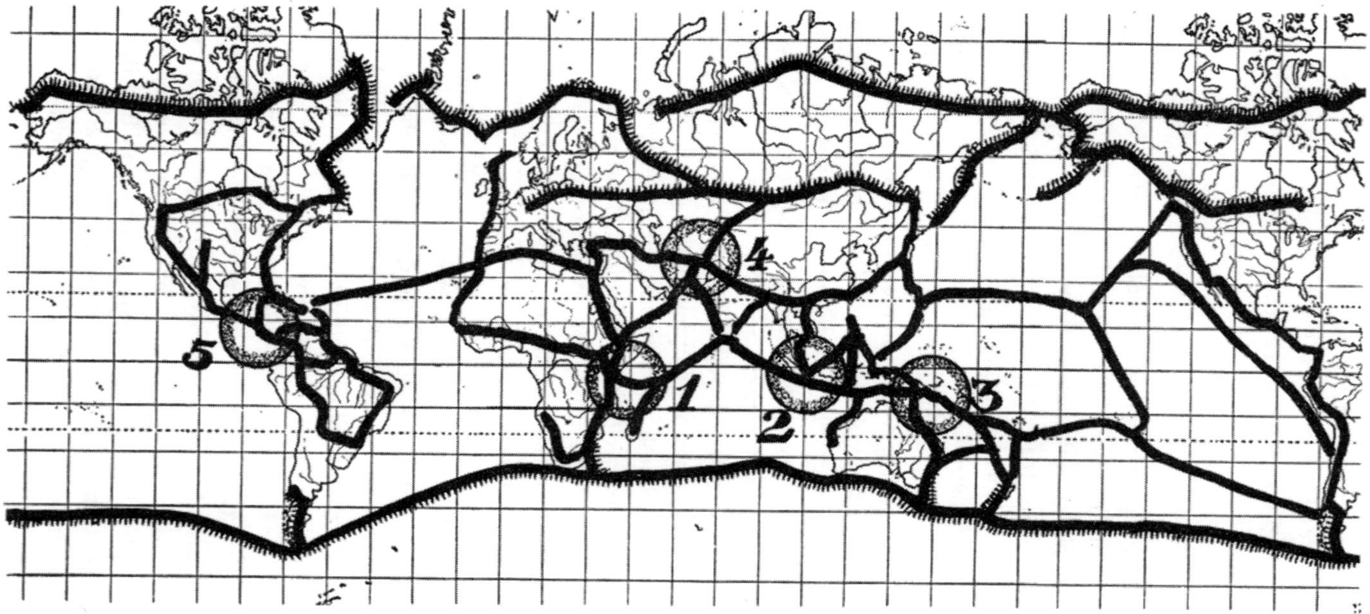

Figure 1.2. Croizat's (1958, IIb: 1018, figure 259) summary of the major features of global biogeography of both plants and animals. The lines are generalized tracks or repeated distributions of organisms. Numbered areas are panbiogeographic nodes, or major intersection points, of the generalized tracks. Hatched lines represent boreal (in the north) and austral (in the south) distribution patterns. [Image courtesy of John R. Grehan.]

Biogeographic patterns are not all necessarily explained by current, generally accepted, well-known geological hypotheses or familiar details of plate tectonic theory. Many biogeographers have long called for the recognition of a formerly closed Pacific basin to explain the distribution of its life (Chapter 8). This theory is still controversial, and many geologists reject the notion of a closed Pacific basin as folly. But more data, both biological and geological, may change this, just as Wegener's notion eventually changed the accepted early-20th-century paradigm of Earth history. Seen until now as part of a widening rift in biology, the interdisciplinary approaches of systematics and evolutionary biology are united with Earth history under the multidisciplinary comparative biogeographical approach.

ORGANIZATION OF THIS BOOK

The core of this book is organized into three parts:

Part I: History and Homology In Chapters 2, 3, and 4, we detail the foundations of comparative biogeography and explain how they relate to the interconnected fields of systematic and evolutionary biogeography. Endemism, the restriction of organisms to particular places, is introduced as one of the core concepts in biogeography. Our thesis is that discovery of a classification of endemic biotic areas that specifies a pattern of area relationships logically precedes inferences about the mechanisms or processes that may have caused biotic distribution.

Part II: Methods We review current methods of biogeography, especially with regard to how they relate to the goal of biotic area classification, in Chapters 5 (Processes) and 6 (Methods and Applications). Our aim is not to exhaustively critique all biogeographic methods, an activity which would be well beyond the scope of this book, but rather to contrast some of the methods, and especially their assumptions, with those of comparative biogeography. In Chapter 7, we outline our method of systematic biogeography, which is discovering a global biotic area classification.

Part III: Implementation We address the relationship between Earth history (geology) and biological distribution in comparative biogeography in Chapter 8. We then tackle the complex biogeography of the Pacific in Chapter 9 to implement our method, demonstrating the power of biogeography to discover and interpret natural patterns.

We close with Chapter 10, our vision for a global biogeography. We argue that biogeography is Big Science and deserves the attention and resources given to other large-scale, global scientific efforts.

NOTES

1. Panbiogeography, formulated by botanist and biogeographer Léon Croizat (see Box 1.2), documents and interprets distribution patterns with respect to each other without relying on or specifying particular phylogenetic hypotheses. We share many basic principles with panbiogeographers, but we differ in the use of phylogenetic patterns in biogeography.

2. The first proposal of an ancient supercontinent, Pangea, is often credited to the 18th-century French naturalist Georges-Louis Leclerc, Comte de Buffon (1766). In contrast, Papavero et al. (2003) argue that Buffon borrowed the idea from German scholar and collector Johann Wilhelm Karl Adolph von Honvlez-Ardenn, Baron von Hüpsch (1764), who published it two years earlier.

3. Darwin, as well as Wallace, fell under the influence of Lyell. Craw (1984:49) argues that "Darwin, in his first 'Transmutation of Species' notebook (1837–1838) used biogeographic evidence to erect novel geological hypotheses. These included a continental drift theory in which all the continents were grouped together into the middle of the Pacific Ocean. . . . Subsequently in his 'On the Origin of Species' (1859) he rejected that view and argued vehemently in favour of the permanence of continents and oceans. . . . In his mature work particular geological theories were used as the basis upon which biogeographic narratives were constructed."

4. We see the field of biogeography as logically integrative because it combines biology, ecology, geology, geography, paleontology, and so on. All biogeography is "integrative biogeography," and this view has a well-established historical foundation. The phrase "integrative biogeography" has been used to endorse a particular set of methods (sensu Donoghue and Moore, 2003) or to inflate artificial divisions, such as that between phylogeny and ecology (Wiens and Donoghue, 2004; see Chapter 5).

FURTHER READING

Fortey, R. 1996. *Life: A natural history of the first four billion years of life on Earth.* Alfred Knopf, New York.

Knoll, A. H. 2003. *Life on a young planet: The first three billion years of evolution on Earth.* Princeton University Press, Princeton, New Jersey.

Leviton, A. E., and M. L. Aldrich (eds.). 1986. *Plate Tectonics and Biogeography: Earth Science History, Journal of the History of the Earth Sciences Society*, 4(2) [1985] 91–196.

Nield, T. 2007. *Supercontinent: Ten billion years in the life of our planet.* Harvard University Press, Cambridge, Massachusetts.

PART I

HISTORY AND HOMOLOGY

TWO

HISTORY AND DEVELOPMENT OF COMPARATIVE BIOGEOGRAPHY

THE MEANING OF PLACE

Biogeography has a history as long as that of biology. From the earliest times, places were identified by their plants and animals. As organisms were classified, so were places; classification of organisms was naturally intertwined with classification of areas. Inevitably, development of and interest in the theory and methods of area classification have followed closely those of organismal classification. We briefly review that history as it relates to the development of comparative biogeography.

Overview

Biogeography is the study of the natural relationships between organisms and the places they live. The study of biogeography is as old as the study of biology itself.

Classifications of areas have often been depicted on maps, although the meaning and use of these maps has been controversial.

Interactions between the biotic and abiotic parts of the world are complex. Natural classifications make sense of this complexity by hypothesizing relationships among biotic areas. Explanatory mechanisms may be added to interpret this complexity.

We introduce here the notion of area relationship, known as *area homology*, which forms the basis of geographical congruence or area monophyly. An *area homolog* is a statement of relationship among three or more areas.

Without natural classifications of areas, biogeographers cannot compare area relationships. The search for natural biotic area classifications is the foundation of comparative biogeography.

Early naturalists pondered both the diversity and the distribution of plants and animals. In his *Historia Animalium*,[1] Aristotle identified animals with the places they lived: "There is a bird that lives on rocks, called the blue-bird from its colour . . ." (*Historia Animalium,* IX, 21), and the times they lived, "The oriole is yellow all over; it is not visible during winter, but puts in an appearance at the time of the summer solstice, and departs again at the rising of Arcturus [*Alpha Bootis*—among the brightest stars in the Northern hemisphere]" (*Historia Animalium,* IX, 22). He observed that animals are more common in some areas than in others: ". . . lions are more numerous in Libya . . . the leopard is more abundant in Asia Minor, and is not found in Europe at all" (*Historia Animalium,* VIII, 28). On seeing the *same* kinds of organisms in *different* places, he noted, "Locality will differentiate habits also; rugged highlands will not produce the same results as the soft lowlands. The animals of the highlands look fiercer and bolder, as is seen in the swine of Mount Athos; for a lowland boar is no match even for a mountain sow" (*Historia Animalium,* VIII, 29). Places, or areas, have their own character, established and reflected by the organisms that live within them: the black ibis is characteristic of the Nile delta, the lion of Africa and India.

Not surprisingly, the earliest formal studies of place were to identify and classify areas, just as the earliest formal studies of plants and animals were to identify and classify organisms. In the 18th century, Swedish botanist and naturalist Carolus Linnaeus, who shared ideas with fellow student, Petrus Artedi (see Box 2.2), devised an artificial or synthetic classification of organisms, a taxonomy (Linnaeus, 1735, 1758). All plants, based on their similar structural and developmental characteristics, were allocated to one taxonomic group. Each plant could be identified and placed into a classification system based on variable features, such as number and arrangement of flower petals. The smallest unit of classification, the species, was given a two-part name or binominal, along with a list of diagnostic characteristics, and was placed in the group to which it thereby belonged. Thanks to Linnaeus, anyone could identify the same kind of organism using its universal name. At the same time, the French naturalist Lamarck was writing one of the largest floral classifications of France: the landmark *Flore Françoise* (Lamarck, 1778), published the year Linnaeus died.

Biogeographical Maps and Classifications

In 1805, fellow French naturalist A. P. de Candolle joined forces with Lamarck to write the third edition of the *Flore Française,* which introduced a map depicting the floristic areas of France. Flora, like fauna, form recognizable areas with particular, distinctive characteristics. New was the concept of a classification of areas in the form of a map

Box 2.1 Parallel Terms in Comparative Biology

Systematics	Biogeography
Organism	Place
Taxon	Endemic area
Taxa	Biotic area

A place is characterized by the organisms that live within it over time. A place, like an organism, may be defined by the *list of characteristics* that we use to classify it as an *endemic area* or by the *relationships* it shares with other areas in a biota. A biotic area may refer to any group of endemic areas, be they regions, realms, or the entire world. Within a biogeographic classification, *place* is the geographical equivalent of *organism* in a systematic classification.

Box 2.2 Carolus Linnaeus (1707–1778) and Petrus Artedi (1705–1735)

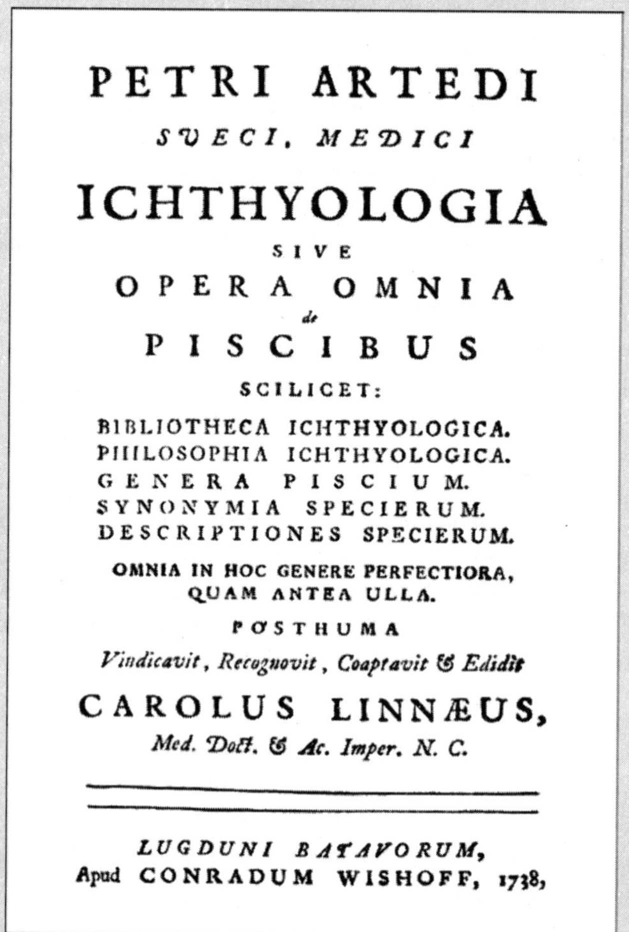

PETRI ARTEDI

SUECI, MEDICI

ICHTHYOLOGIA

SIVE

OPERA OMNIA

de

PISCIBUS

SCILICET:

BIBLIOTHECA ICHTHYOLOGICA.
PHILOSOPHIA ICHTHYOLOGICA.
GENERA PISCIUM.
SYNONYMIA SPECIERUM.
DESCRIPTIONES SPECIERUM.

OMNIA IN HOC GENERE PERFECTIORA,
QUAM ANTEA ULLA.

POSTHUMA

Vindicavit, Recognovit, Coaptavit & Edidit

CAROLUS LINNÆUS,

Med. Doct. & Ac. Imper. N. C.

LUGDUNI BATAVORUM,
Apud CONRADUM WISHOFF, 1738,

Figure 2.1. Title page of Artedi's (1738) *Ichthyologia,* published posthumously by Linnaeus [image courtesy of Theodore Pietsch].

Carolus Linnaeus is credited with devising the first standardized naming system for plants and animals. In his *Systema Naturae,* Linnaeus proposed a hierarchical classification system that grouped species and genera into classes and orders based on an arbitrary quantitative system (e.g., grouping plants based on the number of their stamens). Petrus Artedi, the father of modern ichthyology, was Linnaeus's close colleague and fellow student studying the natural history collections of Albertus Seba: Artedi's *Ichthyologia* was published posthumously by Linnaeus (Artedi, 1738). In the first edition of the *Systema Naturae,* Linnaeus (1735) adopted the classification of fish as proposed by Artedi (see Pietsch, 1995).

Although a basic nomenclature and taxonomy had existed since Aristotle, Linnaeus's system standardized names within a rigid classification. Linnaeus considered species to be real and the higher taxa to be artificial, a concept that was later challenged by Johan Wolfgang von Goethe and other 19th-century morphologists. Linnaeus's system of classification met swift criticism, especially from Buffon. During the late 18th to early 19th centuries, taxonomic innovation came from French naturalists Jean-Baptiste Pierre Antoine de Monet, Chevalier de Lamarck, and, in particular, Antoine Laurent de Jussieu, who rejected the Linnaean artificial quantitative system because it grouped unrelated species. Their influence on biological classification was profound.

(Figure 2.2).[2] Maps—pictorial representations of the distribution of organisms—were introduced as a way to convey area classifications.

Lamarck and Candolle's goal was to create a botanical map of France "... designed to highlight two very different things: 1. the knowledge of vegetation in different parts of France that are known by Botanists and; 2. the general plant distribution on French soil." Most important, the map "... should be considered more of an attempt to apply a specific methodology rather than an attempt to show the complete plant geography of France" (Lamarck and Candolle, 1805; translation from Ebach and Goujet, 2006:763). This goal, to depict or map French floristic provinces, gave rise to an area classification of the world (Candolle, 1820; see below and Box 2.3).

Candolle's largest and most lasting contribution was to recognize that the areas in which an organism lives could be ranked. He called the smallest area a *station:* "the special nature of the locality in which each species grows" (Candolle, 1820, in Nelson, 1978:280). By *station,* he meant

Box 2.3 A Comparison Between Biogeographical and Geographical Distribution Maps

Biogeographical Map	Geographical Distribution Map
Aims to depict biotic area classifications (e.g., areas of endemism, regions, etc.)	Aims to show taxic distributions and distributional pathways (e.g., centers of origin, dispersal routes, etc.)
Mechanism-independent classification that is based on the organisms that characterize a place	Dependent on evolutionary mechanisms (e.g., sympatry, inferred dispersal routes, etc.)
Depicts natural endemic biotic area boundaries	No biotic area boundaries shown
Not based explicitly on, geopolitical cultural, or geographical regions or boundaries	May be based explicitly on geopolitical, cultural, or geographical regions or boundaries
Lamarck & Candolle (1805) produced the first known biogeographical map (see Ebach and Goujet, 2006)	Zimmermann (1777) produced the first known geographical distribution map (see Camerini, 1993)

German zoologist and geographer Eberhard August Wilhelm von Zimmermann (1777) revolutionized taxonomy by presenting a geographical distribution map. The writings of Pliny the Elder, Buffon, and Linnaeus inspired Zimmermann to draw a map of mammals *(Tabula Zoographica)* to depict their distribution in Asia, Europe, and North and South America. Zimmermann's map, following Aristotle and Buffon, was strictly distributional, showing the most southern and northern latitudinal migratory/distributional ranges of several species (e.g., elephants, camels), rather than proposing and classifying biotic regions (e.g., Figure 2.5). Zimmermann's work represented a major step forward in biogeography as it represented distributions visually; until then, they had only been represented verbally. Biogeographical maps were added to zoogeographical and phytogeographical studies in the early 19th century. Maps that show biogeographical regions may be considered *biogeographical,* and those that show distributions or distributional pathways *geographical* (see also Camerini, 1993).

what we now call the habitat: the place to which the organism is suited. The *station* or habitat is not fixed; it may change. The Koaro, *Galaxias brevipinnis,* is a bony fish that lives and breeds in the freshwater streams of south and southeastern Australia (including Tasmania); New Zealand; and the Chatham, Auckland, and Campbell islands (Figure 2.3). The Koaro is migratory: larvae are carried passively downstream to marine habitats and return upstream after a five- to six-month period of transformation (McDowall and Fulton, 1996). In the hilly sandstone regions of New South Wales, one of many Koaro habitats, streams and rivers are characterized by deep, eroded gullies. Some streams end in gorges that become ponds that dry up; one of the many habitats available to the Koaro thus disappears. The current distribution of the Koaro throughout southeastern Australia is disjunct (Figure 2.3). The area that encompasses all such habitats was termed the *habitation* by Candolle, meaning "... a general indication of the country wherein the [organism] is native"

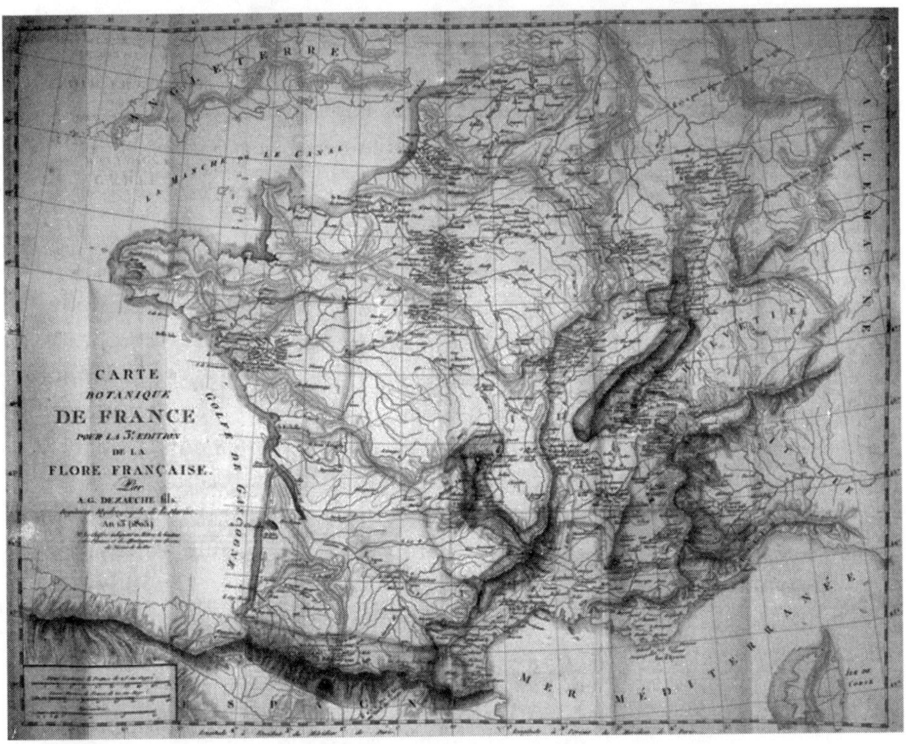

Figure 2.2. Botanical map of France from the 3rd edition of the *Flore Française* (Lamarck and Candolle, 1805). [Image courtesy of Erin Clements Rushing and the Smithsonian Institution Libraries.]

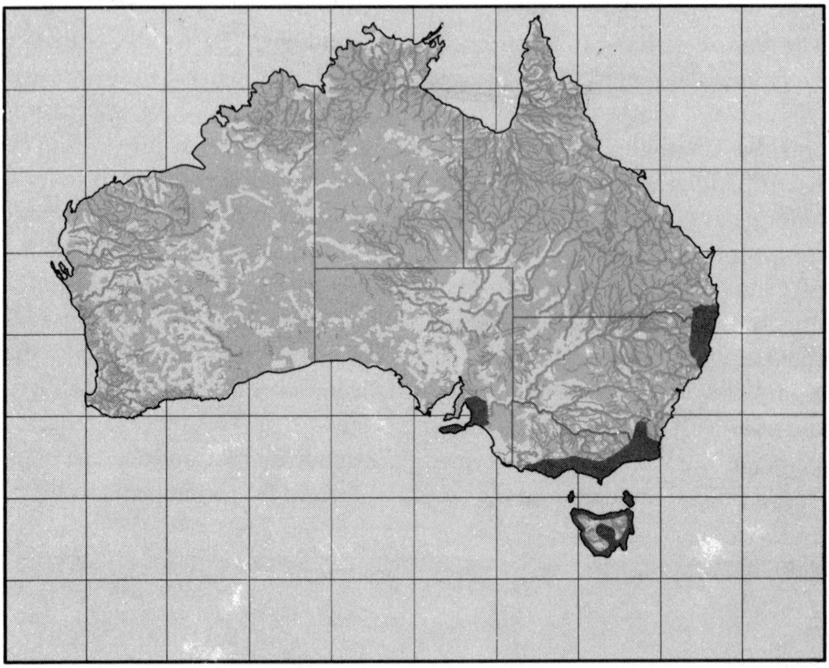

Figure 2.3. Approximate, disjunct distributional limits (shown in dark green) of the Koaro, *Galaxias brevipinnis* throughout southeastern Australia, including Tasmania (after McDowall and Fulton, 1996:54).

(Candolle, 1820, in Nelson, 1978:280).[3] The term *habitation* has been translated into English as "region." A region is a larger area that includes many habitats or stations.[4] For the Koaro, the region includes southeastern Australia, an area that it shares with many unrelated species that do not live in the same habitat (freshwater streams and coastal marine zones), such as the grey ironbark tree, *Eucalyptus paniculata.*

Candolle identified 20 global botanical regions (Candolle, 1820; Nelson, 1978:283–284; Table 2.3), but he was not the only naturalist to classify regions of the world. Many others made similar attempts throughout the 19th century (Tables 2.4 and 2.5). Ornithologist Philip Lutley Sclater (1858) divided the terrestrial world into a minimal six faunal regions. His goal was a classification that would reveal ". . . the most natural primary ontological divisions of the earth's surface" (Sclater, 1858:130). His areas formed a simple division: the old *(Creatio Palaeogeana)* and the new *(Creatio Neogeana)* "creations" or worlds (Figure 2.4).

Despite general agreement between classification schemes of terrestrial plant and animal regions (Table 2.4), the regions are incomplete

TABLE 2.3 CANDOLLE'S (1820) 20 GLOBAL BOTANICAL REGIONS

1. Boreal Asia, Europe, and America	11. Tropical West Africa
2. Europe south of the Boreal region and north of the Mediterranean	12. Canary Islands
3. Siberia	13. Northern United States
4. Mediterranean Sea	14. Northeast coast of North America
5. Eastern Europe to the Black and Caspian Seas	15. The Antilles
6. India	16. Mexico
7. China, Indochina, and Japan	17. Tropical America
8. Australia	18. Chile
9. South Africa	19. Southern Brazil and Argentina
10. East Africa	20. Tierra del Fuego

summaries of global distribution patterns as they omit the more than two-thirds of the Earth's surface that is covered by water; Candolle included only one marine-named region, the Mediterranean, which was focused on the land, not the sea. Ludwig Karl Schmarda, in his *Die Geographische Verbreitung der Thiere* (1853), uniquely included marine as well as terrestrial regions in an area classification (Table 2.5). The marine regions were ignored by his terrestrially focused contemporaries: Sclater excluded the entire marine realm as he gave area of the *Orbis Terrarum* as 45 million square miles (Figure 2.4).

British naturalist Edward Forbes (1846, 1854) focused on the seas and mapped the distribution of marine fishes and the invertebrate molluscs and radiates. He defined 25 provinces that comprised nine horizontal homoiozoic [= homoeozoic, containing similar forms of life] belts (Table 2.5). Andrew Murray's (1866) monumental *The Geographical Distribution of Mammals* included colored distribution maps of both terrestrial and marine mammals. The maps of marine mammals echoed distribution patterns recognized by Forbes: Pantropical (Figure 2.5a) and Antitropical (Figure 2.5b). Furthermore, it was understood that continents did not support homogeneous faunas, even throughout the Pantropical realm. Across the map of South America, Forbes enscribed, "The marine fauna of the two sides of Central & Tropical S. America (almost) wholly distinct as to species." Of Africa, he wrote, "The marine faunas of the two sides of inter-tropical Africa are wholly distinct."

TABLE 2.4 COMPARISON AMONG SOME OF THE TERRESTRIAL BIOGEOGRAPHICAL REGIONS PROPOSED DURING THE 19TH CENTURY

Sclater's (1858) Ornithological Regions	Schmarda's (1853) Zoological Regions	Prichard's (1826) Zoological Regions	Candolle's (1820) Botanical Regions
I. Palaearctic region	I. Polar	1. Arctic	1a. Boreal Europe and Asia
	II. Central Europe	2. Temperate	2. Europe south of the Boreal region and north of the Mediterranean
	I. Polar	1. Arctic	3. Siberia
	V. Mediterranean	2. Temperate	4. Mediterranean Sea
	III. Caspian Steppes		5. Eastern Europe to the Black and Caspian Seas
	VII. Japan		7a. Japan
III. Indian	XIII. India	4. Indian Isles	6. India
	XIV. Sunda Archipelago		7b. China, Indochina
II. Ethiopian	XI. Central Africa	7. Southern Extremities	9. South Africa
	X. Africa Oriental	2. Temperate	10. East Africa
		3. Equatorial	11. Tropical West Africa
			12. Canary Islands

IV. Australian	XV. Australia and	6. Australian	8. Australia
	XXI. Polynesia	5. Papua	
V. Nearctic	I. Polar	1. Arctic	1b. Boreal America
	VIII. North America		13. Northern United States
			14. Northeast coast of North America
VI. Neotropical	XVI. Central America	2. Temperate	15. The Antilles
		3. Equatorial	16. Mexico
	XVII. Brazil		17. Tropical America
	XVII. Peruvian/Chilean		18. Chile
	XIX. Pampas		19. Southern Brazil and Argentina
	XX. Patagonia	7. Southern Extremities	20. Tierra del Fuego

NOTE: Sclater's (1858) six regions are compared with those of Schmarda (1853), Prichard (1826), and Candolle (1820) (see also Nelson and Platnick, 1981; Papavero et al., 1997). Schmarda (1853) also proposed 10 marine regions (Table 2.5).

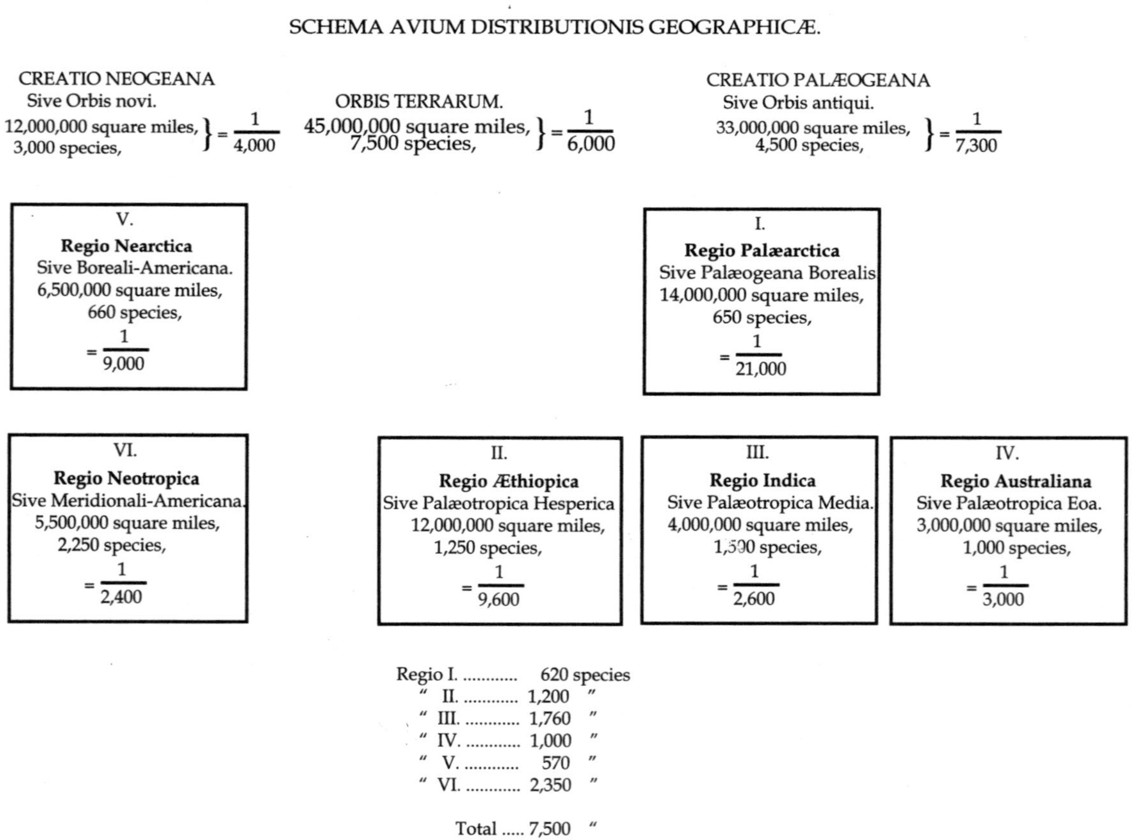

Figure 2.4. Schematic Diagram of the Geographic Distribution of Birds (redrawn from Sclater, 1858:145). Philip L. Sclater's division of the terrestrial world into six faunal regions based on the distribution of birds. The approximate area and number of species living in each region was used to estimate the area occupied by a single species. The neotropics, region VI, is the densest, with one species per 2,400 square miles. The list at the bottom of the figure gives alternate numbers of species per region.

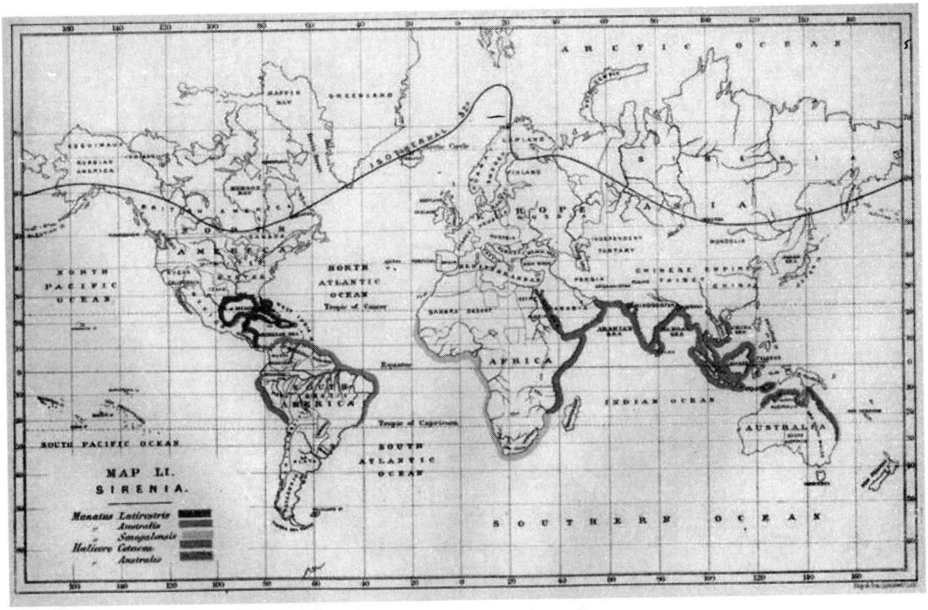

Figure 2.5a. Pantropical distribution of sirenians (aquatic mammals including manatees), as mapped by Murray (1866:198).

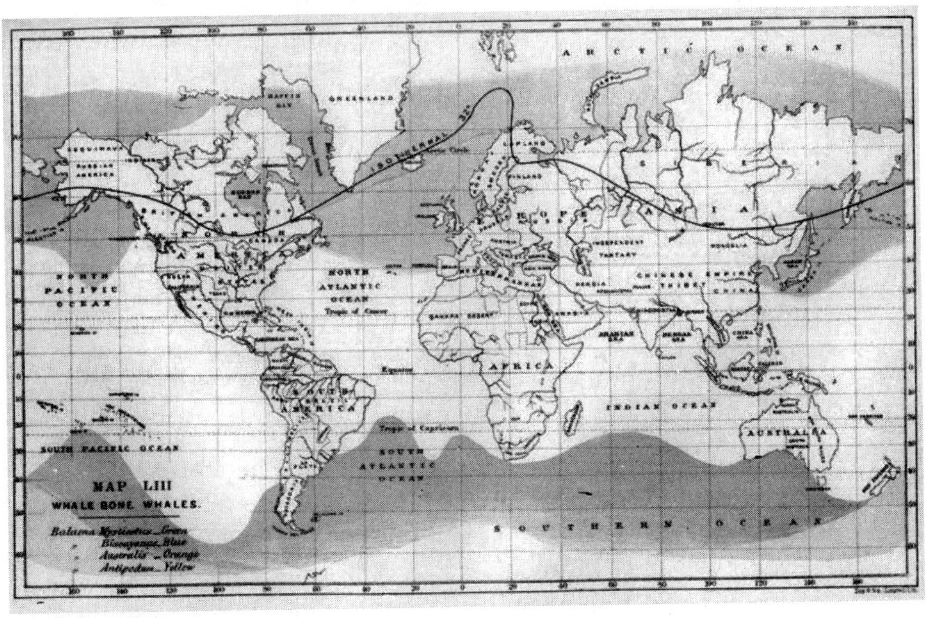

Figure 2.5b. Antitropical distribution of mysticete cetaceans (whalebone or baleen whales), as mapped by Murray (1866:208).

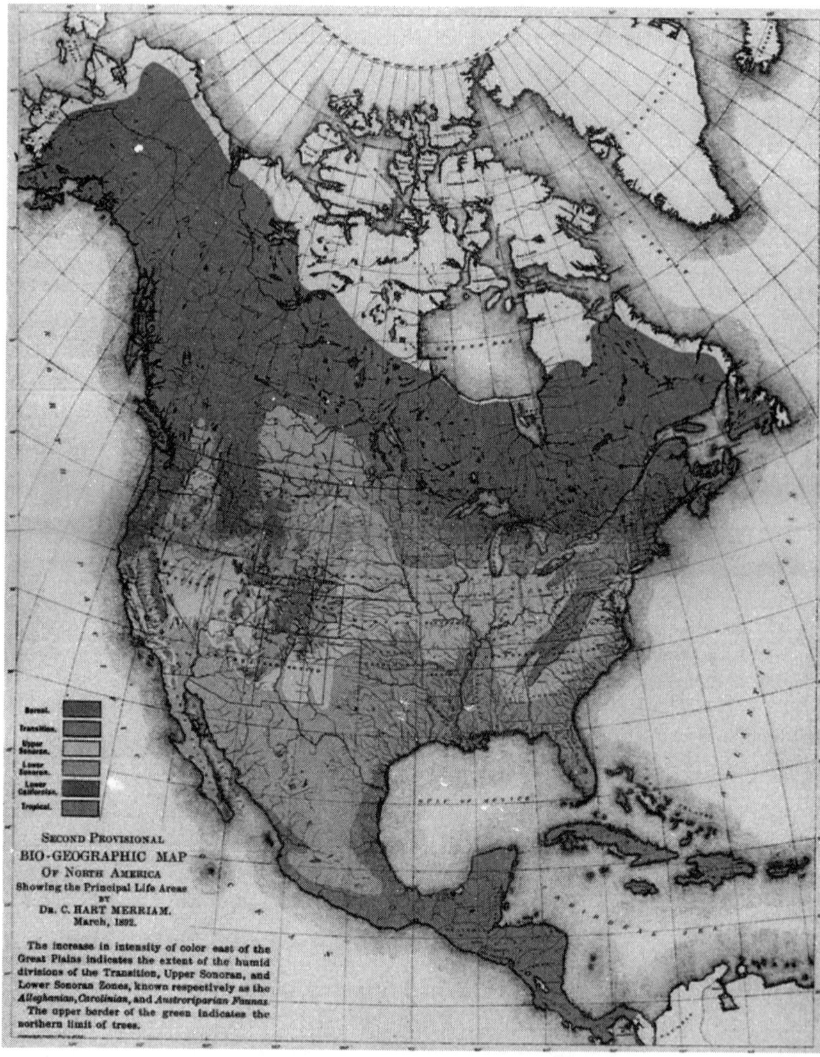

Figure 2.6a. Second Provisional Bio-Geographic Map of North America showing the principal Life Areas (Merriam, 1892). The first use of the word "bio-geographic" in English (see Chapter 1).

Late in the 19th century, American mammalogist Clinton Hart Merriam (1892) proposed an area classification based on *life-zones*, analogous to Candolle's stations. Merriam concentrated on North and Central American terrestrial mammal distributions, which he divided into six life-zones: Boreal, Transitional, Upper Sonoran, Lower Sonoran, Lower Californian, and Tropical. The divisions were illustrated on a colored map (Figure 2.6a) labeled "Bio-geographic," a term used for the first time in English.[5] To

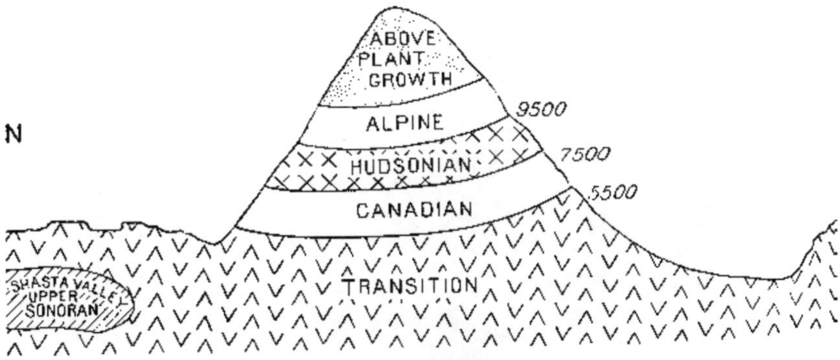

Figure 2.6b. Diagram of Mount Shasta, part of the Cascade Range, showing relative positions of altitudinal life-zones (from Merriam, 1899: Fig. 30).

Merriam, the natural world was established on biogeographical principles. Regions and life-zones were divided by the number of distinct types of organisms rather than by the current topography. A river, sea, or mountain might be assumed to be a barrier, but if the same organisms lived on both sides of the inferred barrier, Merriam argued, then it should not be interpreted as such. Merriam also argued that all regions and life-zones that had particular types of organisms be given equal rank. In his view, a smaller area with a distinctive flora or fauna should not be considered inferior to a larger area that has fewer distinctive types. Unlike Sclater's regions, Merriam's were of unequal size—the colder and less diverse Arctic regions were larger than the highly diverse, warmer southern regions. Merriam also delimited life-zones by altitude (Figure 2.6b).

Attempts to establish a universal classification system for areas—the ontological divisions of the Earth's surface—met with mixed success. The examples noted here are difficult to resolve as a single, consistent system. Because Sclater's regions were larger, they could accommodate the smaller regions of Candolle and of Merriam, but they contradicted those of Forbes. Candolle's regions were characteristic of flora and ignored the smaller regions that distinguished Merriam's life-zones. Merriam, for whom all areas were equal, opposed a hierarchy of areas from smallest to largest. Marine biogeography was seen as patently distinct from terrestrial biogeography. As a result, plant geographers used Candolle's areas, mammalogists and ecologists adopted Merriam's, general zoogeographers preferred Sclater's, and marine zoogeographers adopted Forbes's.

No consensus had been achieved when Wallace (1876), considering all the options available to him, chose Sclater's regions alone. Wallace did not intend to arbitrate on behalf of biogeographical classifications. He wanted "convenience, intelligibility and custom" (Wallace, 1876:54)—

Figure 2.7. The Neotropical Region of Wallace (1876), divided into four subregions: (1) Chilean, (2) Brazilian, (3) Mexican, and (4) Antillean.

that is, a simple classification.[6] Perhaps it was while aiming for simplicity that Wallace (1876) decided to recognize Sclater's Neotropical Zone, even though he divided it into four subregions (Figure 2.7): Chilean, Brazilian, Mexican, and Antillean. The Chilean subregion lies in the Austral (Figure 2.5b), not the Pantropical (Figure 2.5a) realm of Murray,

TABLE 2.5 FORBES'S (1854, 1856) 25 PROVINCES OF MARINE LIFE (FISHES, MOLLUSCS, AND RADIATES), COMPARED WITH SCHMARDA'S (1853) 10 MARINE REGIONS, ARRANGED BY FORBES'S NINE HOMOIOZOIC [= HOMOEOZIC] BELTS

Forbes's Homoiozoic Belts	Forbes's Provinces	Schmarda's Marine Regions
I. North Polar	I. Arctic	XXII Northern Ice Sea
II. North Circumpolar	II. Boreal	XXIV. North Atlantic Ocean
	XII. Ochotzian [= Okhotskian]	XXVI. North Pacific Ocean
	XIII. Sichtian [= NE Pacific]	
III. Northern Neutral	III. Celtic	XXIV. North Atlantic Ocean
	XI. Mantchourian	XXVI. North Pacific Ocean
	XIV. Oregonian	
	XXV. Virginian	
IV. Northern Circumcentral	IV. Lusitanian	XXV. Southern European Mediterranean Sea
	V. Mediterranean	
	X. Japonian	
	XXIV. Carolinian	
	XV. Californian	
V. Central	VI. West African	XXVII. Mid/Central Atlantic Ocean
	VIII. Indo-Pacific	XXVIII. Indian Ocean
	XVI. Panamian	XXIX Mid/Central Pacific Ocean
	XXIII. Caribbean	
VI. Southern Circumcentral	VII. South African	XXX. Southern Atlantic Ocean
	IX. Australian	XXXI. Southern Pacific Ocean
	XVII. Peruvian	
	XXII. Urugavian	
VII. Southern Neutral	XVIII. Araucanian	XXX. Southern Atlantic Ocean
	XXI. East Patagonian	XXXI. Southern Pacific Ocean
VIII. Southern Circumpolar	XIX. Fuegian [= Tierra del Fuego]	XXX. Southern Atlantic Ocean
		XXXI. Southern Pacific Ocean
IX. Southern Polar	XX. Antarctic	XXIII. Antarctic Sea

more in agreement with Forbes's classification of South America than with Sclater's. Wallace's area classification contradicted many known global biological distribution patterns; it was convenient, but not natural.

Should the regions of biogeographical maps represent real, natural areas? All proposed regions reflect the distribution of biota. Yet relationships—floral and faunal, marine and terrestrial—were not explicitly described in the schemes of Candolle, Sclater, Forbes, Merriam, or others of their time. Establishing whether the included biotas form natural groups has been contentious since the areas were proposed.

EXPOSING THE IDEA

Despite what we see now as the system's shortcomings, Sclater's (1858) six regions were readily adopted, and his system was perceived as both a convenient and a natural biogeographic division of land areas of the globe (e.g., Udvardy, 1975; Berra, 2001). The formal, stable, and, since its proposal, largely untested but widely defended classification of Sclater (Figure 2.4) has little in common with Croizat's major features of global biogeography (Figure 1.1). The discovery and description of a natural biogeographic classification is our goal. Which, if either, of these classifications represents the natural or ontological division of the Earth's surface? And how do we decide?

Biogeographical Classification in the 20th Century

Biogeographers of the early 20th century included scholars of divergent experiences and interests: oceanographers, geographers, ecologists, morphologists, and a new breed of biologists who focused on populations. Population biologists, in particular, noted that the arbitrary regional divisions of Sclater, promoted by Wallace, were unnaturally rigid. The regions on these biogeographical maps were not present in nature (apart from, one might argue, abrupt divisions such as Wallace's Line, where one could "see" where parts of the Australian and Asian biotas were adjacent [Camerini, 1993]). In delimiting population and species distributions, limits of regions were fuzzy; like inorganic regions, they changed over time, from season to season, from year to year. Sclater's and Wallace's global divisions were not natural and were practically useless for comparing organisms of different taxonomic hierarchies, not to mention for making comparisons of organisms in the sea, or between those of the land and the sea.

Area classification was brushed off by systematists of the modern evolutionary synthesis (Mayr, 1946; see Chapter 1). Classifications were considered subjective and unrepeatable (see also Ragan, 1998). The act of classifying was seen as lacking consideration of evolutionary mechanisms or processes. Adding assumptions about the process of evolution appeared not only to invigorate and modernize systematics and biogeography: it also allowed precedence to be given to explanatory mechanisms over classification. In the end, it ". . . relegated phylogenetics to a secondary role, if that" (Ragan, 1998:8). A classification could be rejected or altered if a mechanism were thought to more plausibly support another classification.

In biogeography, architects of the modern evolutionary synthesis were following Darwin's lead (Camerini, 1993:718):

> Zoological regions were a necessary step in the development of Darwin's theory. Although he used the concept and a map image of mammalian regions in the early formulation of his argument for a common descent and continued to be interested in regional schemes, the regions themselves faded from his writing in the 1850s. He became more interested in the process of natural selection, in the origin of adaptations, and in explaining the evolutionary significance of certain distributional patterns . . . than in the overall geographic regional patterns that resulted.

Thus, focus on evolutionary processes shifted comparative biology away from classification of both areas and organisms. Dispersal from a center of origin and a biological species concept were assumed *a priori*—hence, they were not subject to test. When patterns of species or population distribution did not fit those assumptions, as they did not for Bahamian populations of the landsnail genus *Cerion* (Mayr and Rosen, 1956:39), biogeographers were stumped:

> We have thus the paradoxical situation that colonies 500 kilometers distant are exceedingly similar while adjacent colonies . . . are quite different. Yet where such different types come in contact, they interbreed freely.

In hindsight, it is not a surprise that there was confusion. Distance between organisms and their ability to interbreed are not the results of the same mechanism (see Croizat, 1958; Rosen, 1979). Conflating distance and potential for interbreeding created an invalid comparison. The question that should be asked is, How are the distribution and relationships among the *Cerion* populations like those of other taxa and how are they different? Biogeography must be comparative to differentiate a distribution that is paradoxical from one that is part of a general pattern. To be comparative, biogeography needs area classification.

Mechanisms and Teleology

Classifications of forms or structures should be derived independently of mechanisms or of inferred purpose. The practice of giving a natural structure, form, or process a purpose to explain it is called *teleology*. Some examples of teleology are comical. French anatomist Bernardin de Saint-Pierre (1804:57–58) believed that the obvious stripes running diagonally along a melon mean that it was ". . . intended to be eaten by a family: there are examples of it even in the Indies, and based on our premises, a pumpkin could be divided and shared with one's neighbors." Historically, teleology is associated with a designer and/or purpose in mind. Many anatomists and taxonomists, therefore, arranged organisms based on a functional purpose. The insectivores, a group of mammals that includes shrews, moles, and hedgehogs, were originally classified together because it was assumed they had been designed to eat insects. Yet many organisms, even humans, eat insects. Insectivory does not form the basis of a natural or real classification. Teleology is, therefore, unscientific.

Alas, some biogeographic models are teleological. Birds fly, and some therefore assume that their distribution patterns must be the result of long-distance dispersal. Here, wings serve to do more than facilitate flight. They symbolize random, long-distance dispersal explanations, even when such explanations may be unnecessary (see Wolfson, 1948). Furthermore, dispersal mechanisms, such as long-distance dispersal via rafting, are scenarios that cannot be proven, tested, or observed (see Chapter 5). A field that is reduced to classifying organisms or areas based on a purpose makes the fundamental flaw of assuming primacy of mechanism over form. It abandons classifying groups based on their natural characteristics and resorts to grouping by narratives. Teleology is a persistent problem in science—biogeography is not unique in dealing with this issue. But biogeography is a field of contradictions, in part because of teleology. Here, we aim to provide some relief from these contradictions by focusing attention on non-teleological biogeography.

CONTRADICTORY BIOGEOGRAPHY

Wallace established a global, continent-based, terrestrial classification for both animals and plants, making zoogeography (distribution of animals) and phytogeography (distribution of plants) redundant. Yet a

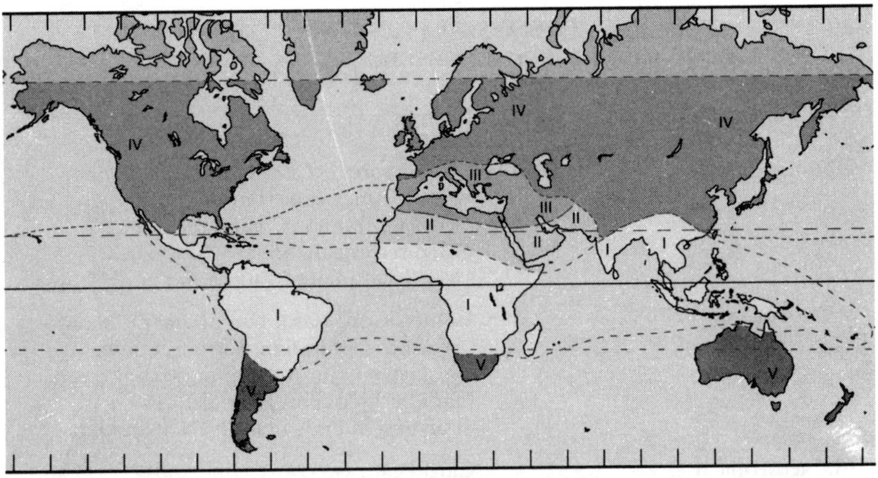

Figure 2.8. The five (I–V) major plant regions of the world, modified from Newbigin (1950: Fig. 33).

relationship between a global classification and the distributional history of a taxon was not acknowledged. Instead, once the center of origin of a taxon was inferred, so too were an ancestor and the starting point of a dispersal pathway. Distributional studies were not seen to require an area classification.

The terms zoogeography and phytogeography were used in the 20th century as if they represented conflicting versions of biogeography. Zoogeography, as practiced by William Diller Matthew, Philip Jackson Darlington Jr., and George Gaylord Simpson, and phytogeography, as practiced by Stanley Adair Cain, Ronald Good, and Armen Takhtajan, among others, were kept distinct because of the *a priori* assumption that animals and plants had different distributional histories. More important for obscuring comparisons, distributions were examined one taxon, plant or animal, at a time.

Takhtajan (1986) echoed Good (1964) in defining six floristic kingdoms (or regions), which were subdivided into a total of 35 regions (or subregions; Table 2.6). Scottish botanist Marion Newbigin (1950; Figure 2.8) had earlier mapped the world's floristic regions, not recognizing a smaller, separate Cape Kingdom of South Africa, which had been included by Takhtajan and others. Botanists, aligned with marine zoogeographers such as Forbes and Murray, recognized a distinct Holarctic region while rejecting a Neotropical Zone as a natural biogeographic region, contrary to Sclater and Wallace.

TABLE 2.6 TAKHTAJAN'S (1986) SIX FLORISTIC KINGDOMS AND THEIR INCLUDED 35 REGIONS

Kingdoms	Regions
I. Holarctic	Circumboreal, Eastern Asiatic, North American Atlantic, Rocky Mountain, Macaronesian, Mediterranean, Saharo-Arabian, Irano-Turanian, Madrean
II. Paleotropical	Guineo-Congolian, Usambara-Zululand, Sudano-Zambezian, Karoo-Namib, St. Helena and Ascension, Madagascan, Indian, Indochinese, Malesian, Fijian, Polynesian Hawaiian, Neocaledonian
III. Neotropical	Caribbean, Guayana Highlands, Amazonian, Brazilian, Andean
IV. South African	Cape
V. Australian	Northeast Australia, Southwest Australian, Central Australian
VI. Antarctic	Fernandezian, Chile-Patagonian, South Subantarctic Islands, Neozeylandic

Swedish marine biologist Sven Ekman's influential view of marine zoogeography was published in German in *Tiergeographie des Meeres* in 1935 and in an expanded, posthumous English version as *Zoogeography of the Seas* in 1953. Ekman defined major zoogeographic regions, such as the Indo-West Pacific, by their geographic boundaries and their distinctive fauna. Antitropical distributions were mapped and their presumed mechanisms discussed. Ekman set the course for other 20th-century marine zoogeographers, particularly mollusc (e.g., Schilder and Schilder, 1938, 1939) and fish (e.g., Briggs, 1974; Springer, 1982) specialists, who continued the search for natural marine regions, especially throughout the Pacific (see Chapter 9).

Despite the agreement between some global biogeographic regions, as above, marine zoogeography has largely stood apart from terrestrial biogeography and phytogeography (e.g., Briggs, 1995; Golikov et al., 1990) because of an emphasis on mechanisms: devising distributional pathways based on assumptions about larval life-spans, generation times, or ocean currents. The marine/terrestrial rift can be mended by focusing first on the identification of repeated patterns, and then on mechanisms (see Chapter 9; Parenti, 2008).

Drawing the Line

One of the most problematic distinctions in biogeography concerns the methods, theories, and models of *ecological biogeography* and those of *historical biogeography*. Ecological biogeography is not exclusively ecological. It includes methods proposed by ecologists, population geneticists, and conservationists working at the level of populations over small areas. Methods of ecological biogeography are *historical* because any distribution pattern, no matter the size or taxonomic level, has a history. By history, we mean those events that are both unobserved and unrepeatable, that happened once and cannot be duplicated under experimental conditions. Phylogeography (Avise, 2000; see also Riddle, 2005; and see Riddle and Hafner, 2004) has effectively exposed the correspondence between historical and ecological biogeography. Phylogeography is an historical method that uses genetic data of populations to make assumptions about the cause of distribution patterns. Had Mayr and Rosen (1956) analyzed phylogenetic relationships among Bimini's *Cerion* land-snail populations, they could have been the first phylogeographers.

The division between ecological and historical biogeography emerged full-force in the mid- to late 1960s, and this rift is still debated (compare Ebach and Morrone, 2005, with Crisci et al., 2006, for example). Until that time, the division of biogeographers was between those who studied phytogeography or zoogeography, or between those who studied marine versus terrestrial organisms. Particularly during the early to mid-19th century, phytogeographers largely examined area endemism, as in the case of Lamarck and Candolle's biogeographical map, the floristic area descriptions of Alexander von Humboldt, and the broad southern hemisphere distribution patterns identified by Joseph Dalton Hooker (1853, 1860). Zoogeographers largely studied distributional pathways, attempting to explain why organisms live where they do and how they got there, a tradition started by French naturalist Buffon (1766; see Nelson, 1978). This was not a rigid difference between botanists and zoologists, as zoogeographers had adopted a descriptive-based biogeography following Sclater (1858).

A particular area classification was considered unique as zoogeographers such as Murray mapped just the distributions of, for example, mammalian taxa. Although zoogeographers and phytogeographers alike concerned themselves with area classifications, many adopted Buffon's evolutionary focus on mechanisms, such as German botanist Adolf Engler (1879, 1882), who proposed general distributional pathways and, in the same breath, individual histories of taxa.[7] By the start of the 20th century, zoogeographers and phytogeographers (as well as paleobiogeographers and anthropologists) were philosophically in step, with each proposing the distributional pathways of their taxa and evolutionary mechanisms that explained diversity as well as area classifications. The division between these two ideas—area classification and evolutionary mechanism—did not necessarily result in a division between practicing biogeographers. Many would continue to combine aspects of both approaches into what German comparative anatomist Ernst Haeckel (1866:287) termed *chorology*, the study of the geographic and topographic spread of organisms away from a center of origin. Today, chorology is viewed by some as equivalent to evolutionary biogeography or simply biogeography (see Williams, 2006; Williams and Ebach, 2008).

One popular ecological biogeographic method is that of island biogeography. Island biogeography was formulated by MacArthur and Wilson (1963, 1967) as a method to predict the number of species that would live in a particular area or community and explain how the community had assembled. It was called island biogeography because the theory first addressed the number of species that would colonize a new, or newly uninhabited, island. The concept of an "island" was later expanded to mean almost any isolated area, such as a mountaintop or desert.

The theory was straightforward and appealing. It included the ways an area could gain new species (immigration) and lose species (extinction). The numbers and kinds of species present would depend, above all, on the size of the island and its distance from the mainland. Other factors included competition among species and the length of time the island had been isolated. Another prediction for the theory was that, over time, the population of an island would reach equilibrium, with the number of immigrant species offsetting the number of extirpated species.

Island biogeography is exclusively ecological and focuses on numbers of species in an area. As formulated by MacArthur and Wilson, it requires no phylogenetic analysis. Also, it is explicitly dispersalist, assuming that islands must be colonized from a mainland or source population. It is a trivial conclusion that volcanic islands must be colonized. The question is, What are the relationships among members of the islands' biota? Are they random or do they specify a pattern?[8]

Without a classification of areas, biogeography mirrored taxonomy: evolutionary biologists concerned themselves less and less with classification (see Felsenstein, 2003; Ghiselin, 2006). Natural groups were considered redundant when the ancestor and lineage of an organism

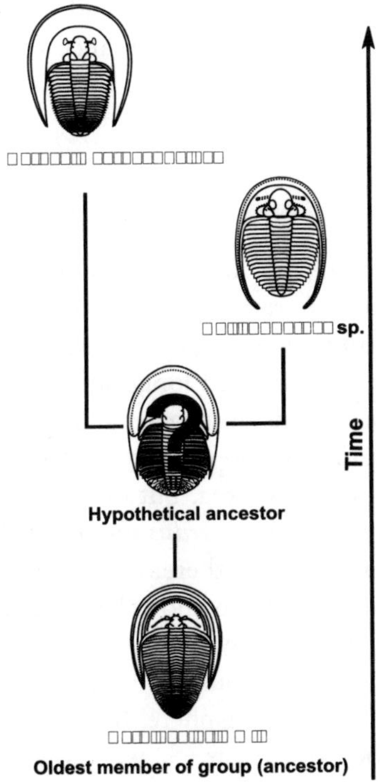

Figure 2.9. Ancestor-descendant gradistics, using the example of trilobites. Under the ancestor-descendant paradigm, taxa are assumed to evolve into each other as inferred from morphological similarity and relative stratigraphic position. The oldest trilobite taxon, *Harpides grimmi,* is the designated ancestor of the group because of its stratigraphic position. A younger hypothetical ancestor with two descendant taxa, *Dolichoharpes* sp. and *Harpes macrocephalus,* is inferred. [Trilobite images courtesy of Sam Gon III.]

were thought to be known. Groups of organisms that shared similar features were considered most likely to be related. One test for this was the stratigraphic record, which gave phylogeny an apparent timeline (Figure 2.9). The oldest member of a taxonomic group, considered the ancestor, was inferred to live in the ancestral area or center of origin. Biogeography and taxonomy were intricately linked by historical scenarios or models, rather than by classifications. Even today, many evolutionary biologists question the point of static classifications that exclude the notion of ancestor-descendant relationships, and others remain skeptical of the value of any classification.[9]

Thus, the real conceptual split we see among biogeographers is not between ecologists and historians, but rather between those who aim to propose a natural classification of areas and taxa and those who propose distributional pathways and infer ancestor-descendant relationships (see MacArthur and Wilson, 1967; Brooks and McLennan, 1991; Crisci et al., 2006).[10]

Recognizing Natural Areas

Areas of endemism (a concept we explore more fully in Chapter 3) are the natural units of comparative biogeography. Areas may be of two general kinds: biotic and abiotic. An area definition that relies solely on abiotic features—say, latitude and longitude—is static. An area definition that includes descriptions of the past and present distributions of its included taxa is dynamic. We do not rely on an arbitrary list of characteristics to accept or reject an area, but on its well-supported taxa and their relationships. A blackbird *(Turdus merula)* afflicted with leucism or plumage discoloration, which is therefore white, not black, may not fit our taxonomy, but the bird's form and behavior are familiar and are recognized by us almost immediately. The same is true for areas. If we wake up on a cold, rocky shore surrounded by penguins, we can be certain that we are in the southern hemisphere. If we walk through the English countryside into a troop of kangaroos, we know something is not right.

Many area definitions use current species or population distributions or the natural barriers that limit them, or arbitrary divisions, such as latitude and longitude, geopolitical boundaries, and so on. As there have been no formal ways to describe natural areas, and given that, over time, area limits vary due to changes in climate, geography, and distribution, many have dismissed biogeography as irrelevant to modern evolutionary studies. Biotic areas, the elements to be classified in biogeography, are organic yet tied to the geography, soil chemistry, and topography of their inorganic environment: "The distribution of organisms is not *assignable* to a . . . geographical space, it *is* the geographical space" (italics in original; Smith, 1989:783; see also Goldenfeld and Woese, 2007). How do we recognize these areas? To do so, we must agree on the definition of a natural area.

AREA HOMOLOGY

A concept of homology in biogeography developed from the work of Léon Croizat (see Chapter 1). Frustration over the lack of a coherent method of comparison of common distributional patterns led him to propose an element of area relationship. That element, area homology, relates areas that share a distributional history through space and time. Area homologies, therefore, depict a *biotic* relationship between areas. With area homologies, biotic areas can be compared to hypothesized geological history over space and time. Area homology differs from the concept of spatial homology *sensu* Croizat (1952): the spatio-temporal relationship among areas considered independently of biotic relationships.

An example may illustrate this important means of comparison. The western and eastern coasts of the northern Atlantic Ocean share similar biota in similar areas. Catadromous eels of the genus *Anguilla* live in streams on each coast and migrate to the Sargasso Sea to spawn. The geological composition of areas is the same insofar as they share a geological history. We find similar rock types, soil types, and fossils in North America and Europe. Today these areas are separated by the Atlantic Ocean. Using Croizat's (1952) spatial homology, we may draw a line, or baseline, to unite these two areas and propose that they were previously connected, either by land bridges or as formerly joined continents; they share a spatiotemporal relationship.

The kinds of organisms that live in the North Atlantic are correlated with the composition and dynamics of the area: soil types, climate, geology, topography, erosion, and so on. The areas, therefore, can tell us much about the life history of the organisms that live there, their physiological requirements, and even their annual cycles. But the areas cannot tell us what kind of organisms these are—their taxonomy—or to what organisms they may be related.

Taxic Homologs, Homology, and Monophyly

In the 1960s and 1970s, the cladistic revolution in systematics reawakened biogeography. Croizat's method deeply influenced a generation of biogeographers, in particular Gareth Nelson, Norman Platnick, and Donn Rosen, all of whom were curators at the American Museum of Natural History, New York (Croizat et al., 1974; Nelson and Platnick, 1981; Nelson and Rosen, 1981; Parenti, 2006).

Croizat's idea of spatiotemporal relationships of areas based on the similarities of their organic and inorganic elements was compelling, yet incomplete. By ignoring the phylogenetic relationship among biotas, Croizat's spatial homology concept is phenetic (based on similarity) rather than cladistic (based on relationship). Although

Box 2.4 Natural and Artificial Classifications

Comparative biologists relate homologs to classify and compare places and organisms. The relationships (homologies or area homologies) that organisms and places share form hierarchical groups that represent a *natural classification.* An *artificial* or *synthetic classification* is based on characteristics (analogs) that are generated by scientists to categorize places and organisms into groups, based on a function (e.g., insect-eating vertebrates) or on an assumption about ancestry (e.g., reptiles gave rise to birds).

There have been repeated calls to retain paraphyletic groups, based on analogs, because they help in classifying ancestors or in recognizing the degree of differentiation among taxa. Recognition of ancestors, or any other such explanatory mechanism, is not part of a natural classification.

areas may share a spatiotemporal relationship based on their inorganic composition and their similar biota, biotic history cannot be interpreted hierarchically without a notion of phylogenetic relationships of taxa. Cladistic biogeographers (viz. Nelson and Platnick, 1981; Humphries and Parenti, 1986, 1999; Morrone and Carpenter, 1994; Crisci et al., 2006) addressed a set of questions, some of which had been tackled by Croizat, using cladistic relationships: What do the similarities among biotic areas mean? Do the geological and geographical histories of areas inform us of their hierarchical biotic relationships? Does the method produce testable hypotheses? Is it empirical?[11]

Cladistics discovers phylogenetic relationships using an empirical method of comparison. What is being compared? We may compare many different organisms based on their appetite for plants. We may group the plant-eaters separately from the non–plant eaters. This forms a highly subjective classification because, given two individuals of the same species, one may like and the other dislike eating plants. Thus, for a comparison to be valid, the groups compared must share a history: this is called a natural group or clade—a group discovered in nature, not fabricated by the researcher. To identify a natural group, called *monophyletic,* rather than to make an artificial group, a comparative method needs an element of form common to all natural groups. That element is a homolog.

Biotic Area Homologs, Homology, and Monophyly

Biota was coined by Leonard Stejneger (1901:89) as a ". . . term to include both flora and fauna which will not only designate the total of animal and plant life of a given region or period, but also any treatise upon the animals and plants of any geographical area or geological period." A *biotic area* is the geographical space occupied by a biota. It may be delimited by obvious barriers, such as a mountain range or river, or may have no recognizable geographic boundaries yet still be identifiable because the biota live in that one place and not another. A biota, for the purposes of systematic biogeography, is *not* all of the organisms of a given region. Distributional limits of the included taxa mark the limits of a biotic area. Choosing a geographical region, such as an island or country, and listing the taxa that live in that area does not constitute the recognition of a biota as there is no reason to assume that all of the organisms in that area share a history. Biotic areas, when properly recognized as organic areas, form the basis of comparative biogeography.

Homology is a concept central to both phylogeny and biogeography. A structure that is found in an inclusive group of organisms, perhaps in different forms, is a *homolog*. Taxa that do not share homologs, or share characters with members of clades other than their own, are paraphyletic, not monophyletic. A paraphyletic "group" is not real or natural, but is one recognized by a taxonomist to fit into an artificial classification. Recognition of natural or monophyletic groups—clades—is at the core of an empirical method of phylogeny reconstruction. Natural groups may likewise be used to relate portions of a biotic area to recognize area homologs.

Why do we seek *area homology* rather than *area similarity*? Croizat compared biotic similarity between areas with geological similarity. Of the two elements, biotic similarity as described by biologists is perhaps the more stable. Geological elements, by definition, differ from area to area. We can compare the basalts (dark volcanic rocks) of any two areas based on similarities in geochemical composition or petrology, for example. The comparison can do no more than confirm or reject that their origin is similar and perhaps point to a particular source. Biotic similarities are equally uninformative as they are based largely on records of presence (or absence) of taxa in areas.

With cladistic methods, we propose a *biotic area homolog*. Biotic areas, like taxa, have parts that are like those present in other individuals or groups as modified structures. In biotic areas, these parts are the taxa and the areas in which they live. A member of one monophyletic group or clade is homologous to another, just as are their parts. Two biotic areas that share taxa from the same clade also share a *homologous "structure"*: the taxa and the areas in which they live. In any hierarchical classification, we compare two taxa to a third to establish a *homologous relationship*. If we discover that two taxa, A and B, share a wing when compared to a third taxon C, which has a fin, then we may hypothesize that A and B are more closely related to each other than either is to C. The smallest unit of relationship is when two taxa are more closely related to each other than either one is to a third taxon. The same is true for biotic area comparisons. Biotic Area A and Biotic Area B may share two taxa that are more closely related to each other than either is to another taxon in Biotic Area C. This three-area relationship—(AB)C—is the biotic area homolog.

When grouped together, biotic area homologs reveal patterns. In systematics, a proposed homolog may relate two taxa that are not related by any other homolog. Rejected as a homolog, it may be considered an analog. Analogs exist among biotic areas. They tell us nothing about the relationships between two or more biotic areas. Once we hypothesize that two biotic areas are more closely related to each other than either is to a third, based on more than one biotic area homolog, we have corroborated biotic area homology.

Discovering taxic homology means that our group (clade) is hypothesized to be monophyletic. The same is true for biotic areas. Biotic area homology uncovers biotic area monophyly: a natural grouping and classification of areas. Biotic area homology means that a group of biotic areas are inferred to share a unique history. The consequences of a natural area classification are the rejection of *a priori* biogeographical models (e.g., Pelletier, 1999) and the adoption of an empirical method to discover historical relationships. In such an application, explaining *individual* distributional histories, for example from centers of origin through distributional pathways, is trivial. Instead, greater attention is given to a *systematic* way to describe areas of endemism and biotic areas and to identify the current and former geographical barriers that do and did effect the current biotic distribution.

By incorporating a systematic approach to taxonomy, it is possible to identify monophyletic groups and reject artificial ones. In so doing, taxonomy has acquired precision. The same is true for a systematic approach in biogeography that aims to use natural biota to define natural biotic regions. Biogeography entails testing hypotheses of area homology among different groups of organisms.

ESTABLISHING A COMPARATIVE BIOGEOGRAPHY

Some claim there is a division between those who study marine and those who study terrestrial organisms. Others believe the division is between those who use molecules and gene trees and those who use morphology and species trees. Others see a division based on ecological versus systematic methods or fossil versus living organisms. Some argue that biogeography is about explaining species histories, while others insist it concerns the classification of biotic areas. And then there are those who argue that all of the methods should be unified under one model. Whatever biogeography now constitutes, it is not consistent and does not form a readily recognizable research program.[12]

There are two conceptual approaches to biogeography: an evolutionary (modeling or mechanism approach) and a systematic (natural classification) approach. The former deals with individual taxa, modeling their distributional histories and proposing evolutionary mechanisms (competition, predation, dispersal, vicariance, etc.), whereas the latter concerns the search for area homology and of the classification of area homologs based on biotic relationships (Ebach and Morrone, 2005). We aim to establish a *comparative biogeography*, a method or approach that incorporates systematic biogeography (biotic relationships and their classification and distribution) and evolutionary biogeography (proposal of possible mechanisms responsible for distributions). We focus first on description and application of a systematic biogeography, to demonstrate biotic area homology, and then explore mechanisms or processes that may have given rise to general patterns. Systematic biogeography, the discovery of classification of regions based on biotic area homologies, is a newly emerging method of biogeography that bridges phylogenetic and distribution patterns to answer the question "What lives where and why?" on Earth.[13] Our goal is the same as Sclater's: to discover the natural, ontological divisions of the Earth's surface. Our results will differ as we know at the outset that his regions contradict many of the overlapping, repeated distribution patterns of the Earth's biota: plant and animal, terrestrial and marine, living and fossil.

SUMMARY

- Biogeography is about the comparative study of place and its organisms.
- Biogeography is often confused with chorology or the study of individual taxon histories through time and space.
- Without a comparative biogeography, is impossible to find natural patterns: expressions of area relationships.
- Area homologs are the basis for discovering area homology or area relationships. Area monophyly, also called geographical congruence, is the basis of a natural biotic area classification and the foundation of a comparative biogeography.

NOTES

1. Aristotle quotations are from D'Arcy Wentworth Thompson's (1910) translation, *The History of Animals*.

2. The earliest known distribution map was published by Zimmermann (1777). The map is *geographical*, as it shows the distribution of mammals, not *biogeographical*, as it does not show the regional distribution and classification of biota. Zimmermann's map inspired 19th-century naturalists, such as Wallace, Huxley, and others: "Distribution, is not a century old, and is contained in the '*Specimen Zoologiae Geographicae Quadrupedum Domicilia et Migrationes sistens*,' published, in 1777, by the learned Brunswick Professor, Eberhard Zimmermann, who illustrates his work by what he calls a '*Tabula Zoographica*,' which is the oldest distributional map known to me" (Huxley, 1894). (See also Box 2.3.)

3. The biogeographic terms "habitation" and "station" were coined by Linnaeus (e.g., Egerton, 1984), who did not explicitly distinguish between them (Nelson, 1978:280–281).

4. The terms "region" and "realm" are often used interchangeably, without regard to absolute rank. "Realm" is used most often by paleontologists for the highest ranked—most inclusive—areas (Westermann, 2000:5).

5. The term "allgemeinen biogeographie" or "general biogeography" was defined a year earlier by German geographer Friedrich Ratzel (1891) in his *Anthropogeographie*.

6. Classifications, for Goethe, were artificial divisions of nature made by us, for us, to suit particular needs. His experience with biological classifications was Linnaean taxonomy—an artificial organization of parts and classes—which was ultimately replaced by a natural classification of *whole* organisms and their relationships. Any thing may be reduced to parts. A car, grizzly bear, or a piece of paper, for example, may be divided into equal parts. Only the bear is a natural whole that was not designed or constructed, but *formed* naturally. The bear may appear to be made of parts, but each part *flows* into another. The bear, as

all living things, has features that recall characteristics of other organisms. Bears have teeth like us, but they are sharper, more like a dog's; they have fur, not like that of a mouse, but like that of a panda. The organism is alive and dynamic, and we cannot divide it arbitrarily into pieces without losing something of the whole. Instead, we observe characteristics, such as the forearm, which, in other forms, appears in other kinds of organisms that we call vertebrates. Geological areas are likewise dynamic. Although inorganic, they change over time with rising and falling sea levels, desertification, ecological succession, and so on. Not surprisingly, Goethe had reservations about Linnaeus's taxonomy: "Nature has no system; she has—she is—life and development from an unknown centre toward an unknown periphery. . . . Regarding what botany calls 'genera' (in the usual sense of the word), I have always held it impossible to treat one genus like another. I would say there are genera with a character which is expressed throughout all their species; we can approach them in a rational way. . . . On the other hand, there are characterless genera in which species may become hard to distinguish as they dissolve into endless varieties. If we make a serious attempt to apply the scientific approach to these, we will never reach an end; instead, we will only meet with confusion, for they elude any definitions, any law" (Goethe, 1995:43–44).

7. Engler (1879, 1882) proposed four botanical realms—the Arcto-Tertiary, Neotropical, Paleotropical, and ancient ocean (see Cox, 2001)—the last of which, interestingly, included a southern ocean biota.

8. Many modifications have been made to MacArthur and Wilson's theory during the past four decades. A general dynamic model (GDM) of oceanic island biogeography was proposed by Whittaker et al. (2008) to incorporate changes of islands through geologic time, for example.

9. The PhyloCode (de Queiroz and Gauthier, 1990, 1992; de Queiroz and Cantino, 2001; see also www.ohiou.edu/phylocode) is a proposed system of biological classification that emphasizes ancestor-descendant relationships. The continued need for and interest in biological classification has been questioned (e.g., Felsenstein, 2003), but others (see especially Sanderson, 2005) note that classification continues to generate heated debate among systematists (viz. de Queiroz, 2000; Stuessy, 2000; Forey, 2001).

10. Gareth Nelson examined Candolle (1820; Nelson, 1978) and concluded that he had made an important theoretical division between stations and habitations: the study of stations, or botanical topography, is today's ecological biogeography, and the study of habitations, or botanical geography, is today's historical biogeography. Quoting from Candolle (1820), Nelson and Platnick (1981:365) offered an explanation of the proximal and distal causes of stations and habitations: "Stations are determined uniquely by physical causes actually in operation, and . . . habitations are probably determined in part by geological causes that no longer exist today" (Candolle, 1820:413).

"Geological causes that no longer exist" could be interpreted in a number of ways. Candolle lived when the Earth was thought to be changing slowly over time. Could these geological changes represent sedimentation or erosion of mountains? During Candolle's day, there was a conceptual war in geology between the *Neptunists* (e.g., Abraham Gottlob Werner, Robert Jameson), who opposed any notion of sudden or catastrophic geological events, such as the uplifting of mountains, and the *Vulcanists* (e.g., James Hutton), who believed Earth history to be violent and volcanic. Modern biogeographers benefit from a less dichotomous view of Earth history. Geologists came to an agreement by the end of the 19th century that the Earth withstands dramatic events, some of which we see today (e.g., volcanoes, earthquakes, hurricanes, tornadoes, tsunamis, and so on).

What did scientists mean when they referred to "geological change"? Lamarck and Candolle (1805:viii) stated, ". . . of all the factors that influence the habitat of plants, temperature is without doubt the most essential" (translation from Ebach and Goujet, 2006:767). The effect of climate on flora may be interpreted in different ways. Prior to Lamarck and Candolle (1805), biogeographical maps were drawn as cross-sections or transects of alpine or mountainous areas (e.g., Giraud-Soulavie, 1770–1784; Humboldt and Bonpland, 1805). Orogeny (mountain building) was a geological event that altered the climate as well as the landscape. Flora on mountains changed with temperature and latitude: "But where the temperature is equal to that of these mountains, these alpine plants can, with certain precautions, be cultivated on the lowest plains. Even some of those that grow in the high Alps are found on the coast, and in the same mountains the same plants grow higher up on the southern slopes than on the northern ones" (Lamarck and Candolle, 1805:xi; translation from Ebach and Goujet, 2006:768).

Soil chemistry and hydrology also figured prominently: "In some texts, importance is placed on the chemical nature of soil in which plants grow" (Lamarck and Candolle 1805:xi; translation from Ebach and Goujet, 2006:768). Candolle was explicit about the factors that determined plant distributions:

> I believe that in a given country, such as France, the causes that determine the plant region [habitation] could be reduced to three:
>
> 1. Temperature, as determined by distance from the equator, height above sea level and southern or northerly exposure.
> 2. The mode of watering, which is more or less the quantity of water that reaches the plant. The manner by which water is filtered through the soil and the matter that is dissolved in the water, which may or may not be harmful to the growth of the plant.
> 3. The degree of soil tenacity or mobility" (Lamarck and Candolle, 1805:xii; translation from Ebach and Goujet, 2006:768).

The relationship between fossil plants that lived during earlier climatic periods and the formation of different soil types was known in Candolle's time. Perhaps soil types and the latitudinal climatic differences are the "geological causes that no longer exist" rather than the topographical—altitudinal—differences within the same area. Candolle never fully differentiated his two fields of topographical and geographical botany; his method was about classification.

11. Panbiogeographers (e.g., Craw, 1989) have argued that cladistic biogeographers who rejected panbiogeographic concepts only considered the concepts of tracks, not the three other main panbiogeographic concepts—node, baseline, and main massing—which may have equivalents in comparative biogeography (see Chapters 6 and 7).

12. Commentaries on the identity of the field of biogeography include Nelson (1978) and Ferris (1980) and, more recently, Lieberman (2000), Ebach and Morrone (2005), and Riddle (2005).

13. The related concept of biogeographical homology has also been explored by Morrone (2001a).

FURTHER READING

Browne, J. 1983. *The secular ark: Studies in the history of biogeography.* Yale University Press, New Haven, Connecticut.

Gee, H. 2000. *Deep time: Cladistics, the revolution in evolution.* Fourth Estate, London.

Morrone, J. J. 2001. *Sistemática, biogeografía, evolucíon: Los patrones de la biodiversidad en tiempo-espacio.* Las Prensas de Ciencias, Facultad de Ciencias, Universidad Nacional Autónoma de México, D.F.

Rudwick, M. J. S. 2005. *Bursting the limits of time: The reconstruction of geohistory in the age of revolution.* University of Chicago Press, Illinois.

Wallace, A. R. 1869. *The Malay Archipelago: The land of the orang-utan and the bird of paradise. A narrative of travel, with sketches of man and nature.* Dover Publications, New York [1962 reprint].

Wallace, A. R. 1881. *Island life, or the phenomenon and causes of insular faunas and floras including a revision and attempted solution of the problem of geological climates.* Prometheus Books, New York [1998 reprint].

THREE

BUILDING BLOCKS OF BIOGEOGRAPHY

Endemic Areas and Areas of Endemism

ENDEMISM

The concept of endemism has long played a central role in biogeography and biodiversity investigations.[1] The modern meaning of the term *endemism* in biogeography has been credited to A. P. de Candolle (1820): a taxon is said to be endemic to an area if it lives there and nowhere else. The concept is useful and universal. Kangaroos are endemic to Australia. Piranhas are endemic to South America. Giant pandas are endemic to China. But not to all of Australia, not to all of South America, and not to all of China. Kangaroos, piranhas, and giant pandas live only in areas to which they are tied historically and ecologically. We continually refine descriptions of endemic areas to be biologically, ecologically, geologically, and geographically meaningful: giant pandas are endemic to bamboo forests in the mountainous regions of southern and central

Overview

Endemism is a key concept in biogeography. Endemism links an organism with a place: a taxon is said to be endemic to an area if it lives there and nowhere else.

Endemic areas are the building blocks of biogeography. Relationships among endemic areas form the basis of biogeographical classification.

Area names are confused in biogeography when the same name is given to more than one area, or the same area is given more than one name.

How we define endemic areas is analogous to how we define taxa in systematics. Both require a nomenclature. Biogeographic areas should not be arbitrary or artificial. Instead, they should be natural: defined by the taxa that live in them and the relationships among those taxa.

China. The place to which the giant panda is endemic is its unique signature—practically, part of its diagnosis.

When taxa are found living in areas other than their endemic areas, they are called *exotic* or are referred to by the mechanism of their distribution: *introduced species,* or *introductions,* and *invasive species,* or *invasives.* For these taxa, we need no systematic biogeographic analysis to assess how and why they came to live in a "foreign" place; it is often known and well documented, especially for plants and animals transported for food, for labor, or as pets. It is to explain the distribution of all the other taxa of the world—those living in their natural habitats—that we undertake comparative biogeography.[2]

Darwin used the term *endemic* in two, possibly contradictory, ways: the area in which a species was produced or the area in which a species lives (see Anderson, 1994:454). If a species lives where it was produced (where it evolved), then the meanings are the same; if a species moves out of the area in which it was produced, and may no longer live there, however, then the meanings differ. As discussed in Chapter 2, Darwin's first definition of endemic—where a species was produced—which does not include its entire range, implies dispersal from a center of origin; that is, dispersal is part of the definition of endemic. It also implies that a species is younger than at least part of the area in which it lives. Applying this definition limits the number and kind of hypotheses that we may propose about areas of endemism and their histories. As we emphasize throughout this book, to begin with such a set of assumptions is contrary to our aim to discover biogeographic patterns and then infer their cause.

Some biologists further discriminate between *paleoendemic* and *neoendemic* species (see Cronk, 1992, 1997): paleoendemics are relict species that have become isolated because of the extinction of close relatives, whereas neoendemics are species that have evolved relatively recently, as a result of changes in habitat or through a process such as evolution of polyploidy, and live near their close relatives. These definitions require a phylogenetic hypothesis of relationships among the species and its close relatives as well as a good understanding of the distribution of all included taxa—that is, a biogeographic analysis. Discrimination between a paleoendemic and neoendemic can only follow identification of a biogeographic pattern.

Endemism may be thought of as the opposite of *cosmopolitanism:* living throughout the world on all or nearly all continents or throughout the seas. Cosmopolitanism is reflected in Pangean distributions of a wide array of taxa, including caddisflies and osteoglossomorph fishes, to name just two clades. These clades do not necessarily have continuous Pangean distributions, in part because of episodes of extinction throughout geologic time, but they are recognizably and undeniably widespread. Isolated areas identified by their high degree of endemism may be referred to as refuges or *refugia,* the locations of inferred, once-widespread taxa now restricted to small, discrete areas.[3]

Although endemism is the opposite of cosmopolitanism, endemic areas need not be small; they may be defined at both large and small biogeographic scales. The Pacific Plate was defined as an area of endemism for shorefishes and other marine organisms, many of which live marginally on the plate or in its center, for example (Figure 3.1a; Springer, 1982). Many other marine clades are distributed largely *off* the Pacific Plate (Figure 3.1b), although their distributions have often been reported as "worldwide." The separate geologic history of the drainages on the west coast and in the alpine regions of Italy is reflected in the localized distribution of sister species of the freshwater fish genus *Padogobius* (Miller, 1990; Figure 3.2): *P. nigicans* is endemic to the west coast drainages that flow into the Tyrrhenian Sea; *P. bonelli* is endemic to drainages that flow into the Adriatic Sea.

The term cosmopolitanism has also been used in a restricted sense to mean the hypothesized widespread ancestral distribution of a taxon prior to vicariance and regionalization and the subsequent evolution of provincialism. In systematic biogeography, cosmopolitanism is qualified to mean the hypothesized widespread ancestral distribution

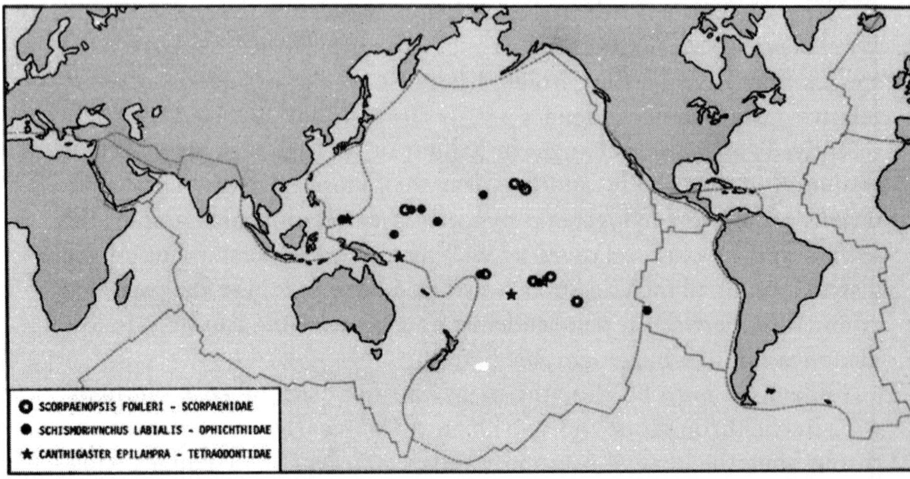

Figure 3.1a. Pacific Plate endemism as demonstrated by three widely distributed marine fish species indicated by symbols, as shown in the key on the left (modified from Springer, 1982: Fig. 26). Plate margins are indicated by dashed lines.

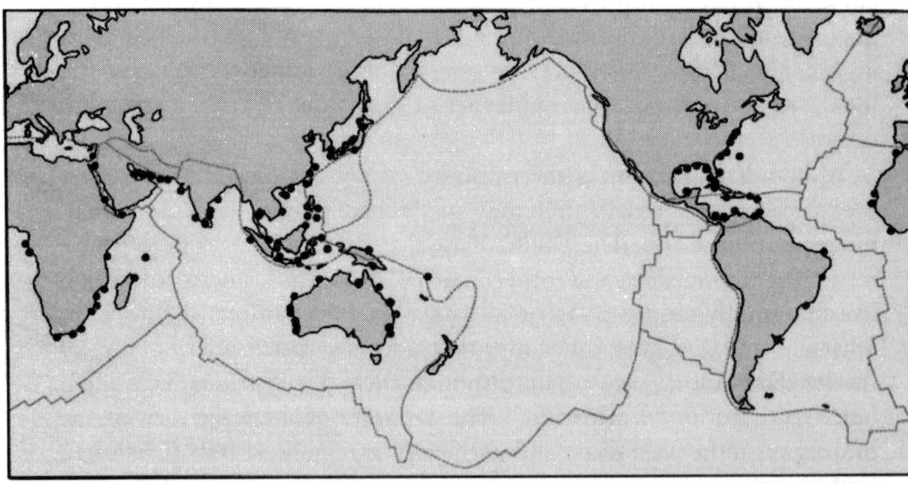

Figure 3.1b. Distribution of the marine fish family Rachycentridae, which is absent from non-marginal portions of the Pacific Plate (modified from Springer, 1982: Fig. 35). Plate margins are indicated by dashed lines.

of a *monophyletic* taxon or clade. For the goby *Padogobius*, this is the area occupied by its included species, *P. nigricans* and *P. bonelli*, as above.

Migration, the regular movement of taxa throughout their life history, is often interpreted as direct evidence for dispersal as a biogeographic

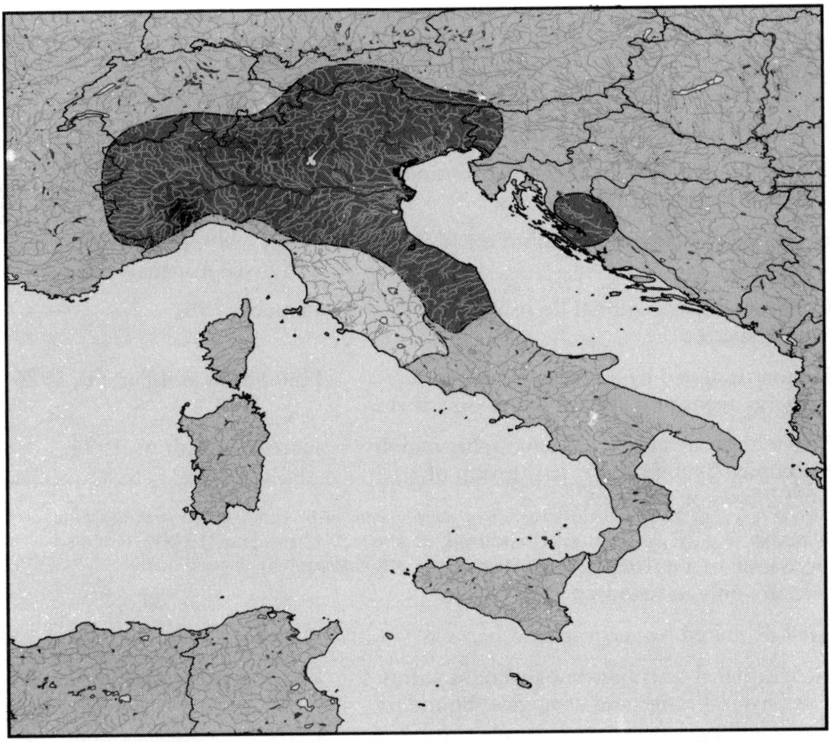

Figure 3.2. Allopatric distribution of Mediterranean gobies of the genus *Padogobius*. *P. nigricans* is restricted to western (Tyrrhenian) drainages (light green), *P. martensii* (called *bonelli* in the text) is restricted to northern and eastern (Adriatic) drainages (dark green). Modified from Miller (1990: Fig. 2).

mechanism. Migration, such as the return to spawning grounds across vast distances, is the natural movement of organisms throughout their range. Darwin (1859) recognized it as the obstinate nature of organisms to return home. Dispersal, the movement (not human-induced) of an organism outside of its natural range, will be defined and discussed in Chapter 5.

Endemic areas are the building blocks of biogeographic analysis. An endemic area is to biogeography what a taxon, such as a species or genus, is to systematics, which emphasizes the systematic nature of bioge ography. Phylogenetic analyses of the taxa that live (or that lived during previous geological epochs) in a set of endemic areas form the data of an area classification. That classification is a hierarchy incorporating increasingly encompassing areas, such as districts, regions, and realms, which form biotic areas.[4]

TABLE 3.1 SELECT DEFINITIONS OF ENDEMIC AREA

Definition	Author
Region to which an organism is particular	Clements, 1905
Area delimited by coincident distributions of taxa that occur nowhere else	Nelson and Platnick, 1981
Delimited or restricted distribution of a single taxon	Hinz, 1989, *sensu* Dansereau, 1957 (see Anderson, 1994)
Congruent distributional limits of two or more species	Platnick, 1991
Region occupied by a monophyletic group of organisms or a species found only there	Humphries and Parenti, 1986
"Area of occurrence": biogeographic region occupied by a monophyletic group of organisms or a species	Harold and Mooi, 1994
A taxon (e.g., a species) is considered endemic to a particular area if it occurs only in that area	Crisp et al., 2001
Area delimited by geographical barriers	Hausdorf, 2002
Geographical distribution of a taxon within its physical range and ecological boundary	Ebach and Humphries, 2002
Recognized by the coincident restriction of two or more taxa	Laffan and Crisp, 2003
An area in which numerous species are endemic	Szumik and Goloboff, 2004
An area containing species not living elsewhere	Domínguez et al., 2006
The smallest area with significantly congruent distributions recognized as significantly different from all other areas at a particular level in nested clade analysis (NCA)	Deo and DeSalle, 2006

Definition of an Endemic Area

Throughout the history of biogeography, definitions of endemic areas have been either solely geological or have combined biology, including paleontology, and geology. The meaning of an endemic area has not always been specified as it was generally thought to be well understood. Application of phylogenetic methods to biogeography has coincided with proposals of explicit definitions of endemic areas (Table 3.1). In systematic biogeography, endemic areas are not inorganic, not defined solely by physical geographic limits, but organic, combining biological with geological parameters—geography, geology, soil chemistry, climate, or other ecological features—by which we may compare different parts of the Earth to each other over time.

An endemic area is *any disjunct or continuous geographical space, through time, that delimits the current and past distribution of one or more taxa*. Even if we agree on this definition of an endemic area, recognizing such an area for biogeographic analysis is more complex: the existence of early Pleistocene fossil giant pandas means that their natural range includes a broad area throughout bamboo forests of southern and central China, as well as localities in Viet Nam and Burma, at least from early Pleistocene times to the present day.

Definition of an Area of Endemism

The concept of an endemic area, the area occupied by a lineage through time, is related to an *area of endemism,* an area whose relationships are being investigated in a biogeographic study. An area of endemism is *the area occupied by at least two purportedly monophyletic taxa*—at least two, because an area occupied by just one taxon will have no history shared with any other area (see Platnick, 1991; Table 3.2). As an example, a genus of ferns may comprise four allopatric species, one each that lives in wet forested regions of southern India, central Madagascar, East Africa and West Africa. Another genus, say of a bird, may have three allopatric species, one each living in roughly the same areas of southern India, central Madagascar, and East Africa. The fern and the bird genera overlap in southern India, central Madagascar, and East Africa, congruent areas of endemism for the two genera about which we may ask, Do the ferns and the birds share a distributional history? Is there a pattern of relationship among these three areas of endemism? West Africa is an endemic area for the species of fern that lives there, but it is not an area of endemism in this hypothetical study because its relationship cannot be part of a pattern that may be shared between the fern and the bird.

The terms *area of endemism* and *endemic area* have often been used interchangeably. *Endemic area* is the more inclusive term.

The concept of endemism has a special role in modern conservation biology. Areas with extremely high numbers of endemic taxa and that have also been degraded by human development may be called biodiversity "hotspots" (Myers et al., 2000). Areas with high numbers

TABLE 3.2 SELECT DEFINITIONS OF AREA OF ENDEMISM

Definition	Author
Regions where populations evolved in isolation	Rosen, 1978
Areas that demonstrate distributional congruence of constituent taxa	Cracraft, 1985
Area occupied by two taxa, with overlapping area identified as a separate area	Axelius, 1991
Smallest coincident ranges of two species and the geographic extent of forest islands	Griswold, 1991
Area defined by the congruent distributional limits of two or more species	Platnick, 1991
Area recognized on the basis of distributions of two or more species	Harold and Mooi, 1994
Smaller generalized tracks	Morrone, 1994
Extensive co-occurrence of *biotic elements* (*sensu* Hausdorf, 2002)	Mast and Nyffeler, 2003; Giokas and Sfenthourakis, 2008
Areas where the distributions of at least two taxa overlap	Quijano-Abril et al., 2006

of species or rare or threatened taxa may also be called "hotspots" (e.g., Possingham and Wilson, 2005). The panbiogeographic concept of a node—the intersection of many distributional tracks—may be thought of as equivalent to a "biodiversity hotspot" (Craw et al., 1999:167). Areas where both endemism and diversity are high are the focus of intense conservation efforts.

The percentage of taxa endemic to an area has been used to characterize the relative endemism of the biota of that area (see Anderson, 1994) and to compare the number of species with size of the area (see Fattorini, 2007). Eighty-nine percent of the approximately 1,000 flowering plants on the Hawaiian archipelago live there and nowhere else, which means that 11 percent of the flowering plants of Hawaii also live elsewhere (Gemmill et al., 2002). We are not concerned primarily with the percentage of endemic taxa in areas insofar as this is a metric of overall similarity (phenetics); it is silent on the phylogenetic relationships of one biotic area to another. Furthermore, as many taxa have undoubtedly undergone serious and significant periods of extinction, the number of species living in an area may simply be interpreted as the number of species that have survived rather than the optimal number that may live in that area.[5]

The study of areas, like the study of taxa, is largely *qualitative*. We stress the qualitative aspects of biogeography, although many biogeographers may use quantitative methods to define and classify areas (see Chapter 6). Establishing definitions and a classification of areas of endemism based on data that ignore area homology and phylogeny—such as relative size of an area, its distance from the nearest mainland, its hypothesized age, its altitude, its depth, and so on—contradicts our principal aim of identifying natural areas that are occupied by natural groups of organisms; it contradicts a systematic biogeography (see also Santos and Amorim, 2007).

TAXONOMIC UNITS AND TAXA

Recognizing and naming endemic areas requires a taxonomy of the organisms living in those areas. The concept of an endemic area may be restated simply as *the place to which a clade is native*.[6] This definition addresses two things:

1. What is a clade?
2. What is the *place*, the geographical space, in which the clade lives?

A clade is the group of plants, animals, fungi, or micro-organisms that one observes and can identify, and, for the purposes of communication, name. To communicate readily, we often first refer to taxa by familiar common names, such as fly, fig, or fish. Each fly belongs to an inclusive taxonomic group, such as a species, which an entomologist diagnoses using the attributes of a single individual or group of individuals, the type specimens, and additional individuals at hand that may be examined to record character variation. Any organism fitting the description will be identified as a member of that taxon. In practice, taxonomic or named groups may be artificial or real. By real, we mean *natural* or *monophyletic*. In one group, we might include organisms that have enlarged and expanded pectoral appendages or wings, such as cicadas, flying fish, and sparrows. The group, even if named formally by a taxonomist, is not real, natural, or monophyletic: it is well supported that sparrows share a phylogenetic history with flightless birds and blue whales that they do not share with flying fish and cicadas. Although this is an obvious example of an unnatural group, in taxonomic practice, many groups are artificial or paraphyletic. The ultimate goal of discovering and recognizing only natural, monophyletic groups (clades) has yet to be achieved.

Figure 3.3a. The Atlantic Salmon, *Salmo salar*. Drawn by Elizabeth Bland, August 1879. [Image from the Illustration Files, Division of Fishes, National Museum of Natural History, Smithsonian Institution; image provided by Lisa Palmer.]

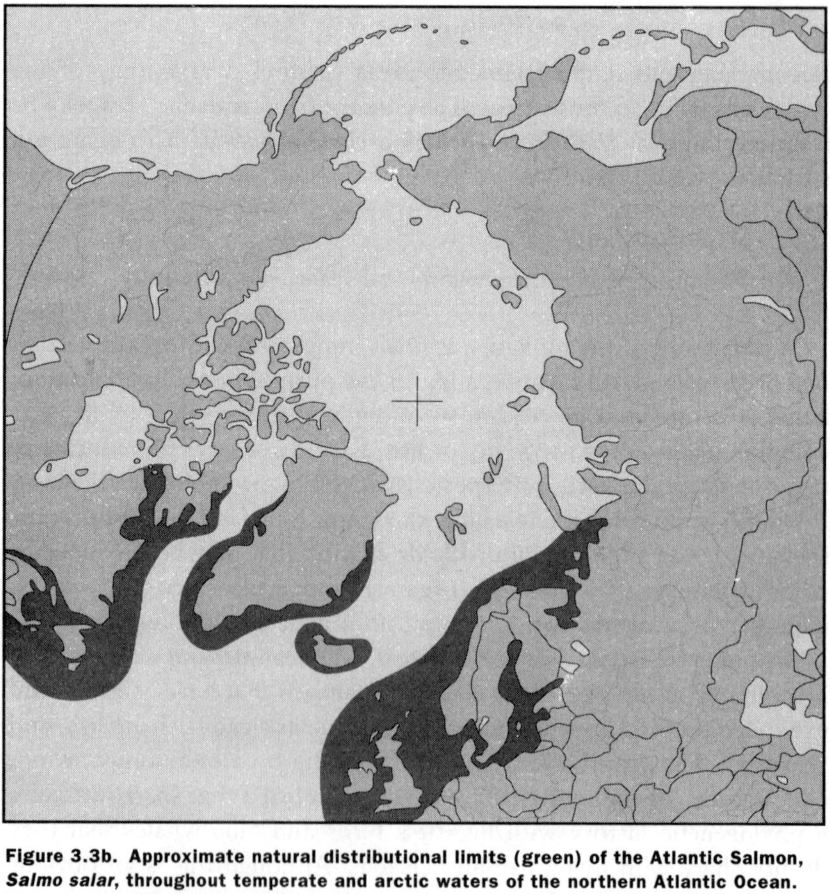

Figure 3.3b. Approximate natural distributional limits (green) of the Atlantic Salmon, *Salmo salar*, throughout temperate and arctic waters of the northern Atlantic Ocean. The cross marks the North Pole.

When Linnaeus and Artedi proposed a taxonomy of life (Chapter 2), it was meant as a way to identify organisms in distinct—be they artificial—groups with a universal name. Taxonomy follows rules that govern the naming of organisms.[7] "Saumon atlantique," "Atlantic Salmon," "lachs," "salmons," "braddan," "bratan," "laks" and "lax," are all common names for the same species, *Salmo salar* (Linnaeus, 1758; Figure 3.3a) that lives throughout temperate and arctic waters of the northern Atlantic Ocean (Figure 3.3b). By standardizing the names of organisms and the larger groups to which they belong, we adopt a nomenclature that biologists may use to communicate clearly and unambiguously about groups of organisms.

Organisms may be both named, taxonomic units with an ordered taxonomy and elements of a natural group. These two concepts need not conflict, just as the names we give to objects do not conflict with the actual objects. In some cases, the name and definition are not sufficient for communication. The most common examples are when the same group of organisms is given two conflicting definitions and names (synonomy) or when the same name is given to two different sets of organisms (homonymy), giving an artificial taxonomy.

We can interpret organisms either as taxonomic units, such as species, which are based on the definitions of individual types and may fall into artificial or real categories, or as elements that share homologous relationships with one another. Given that the study of biogeography and systematics are analogous, there are correspondingly two ways to identify endemic areas: artificial and natural. Once uncovered, artificial areas should be replaced in a systematic biogeographic analysis by natural areas (see below). Interpreting biogeographic history based on the distributions of unnatural or paraphyletic taxa can only lead an investigation toward imprecise or contradictory conclusions (Santos and Amorim, 2007).

TAXONOMIC AREAS AND BIOTA

How should areas be defined? If we look to taxonomy, we see that this is not necessarily a straightforward question. There are more than two dozen species concepts (Wilkins, 2008), for example; some conflict, in part, and others are incompatible, but all are based on a description that is regulated by a nomenclature. A trilobite, an extinct arthropod, in most fossils has only its hard exoskeleton preserved. To describe a trilobite taxonomic unit (at the species level, for example), a paleontologist uses a particular concept—stratigraphic unit—and physical evidence to describe the organisms—the exoskeleton. In the study of living fishes, the process of diagnosis and description is different insofar as the evidence is different. Soft parts, such as muscles, blood vessels, and nerves, or molecular data may be used in addition to the hard parts—the skeleton—to describe taxa. The type of organism the systematist studies invites a different process of identification, description, and diagnosis. This is true also for identifying a place in biogeography.

Places have attracted a range of definitions. Paleobiogeographers (e.g., Jablonski et al., 1985; Lees et al., 2002) may describe place based on geological evidence, such as structure of sedimentary rocks (which provides evidence for a paleoenvironment), geochemistry (evidence for paleoclimates), or fossil distributions. Phylogeographers (e.g., Avise, 2000; Riddle, 2005), who study population gene trees, may use soil chemistry, ecology, and population distributions to define areas. As in taxonomy, each place is defined as an endemic area based on different criteria, and each name is a nomenclatural unit. Nomenclature is blind to the species or area concepts used to describe and diagnose taxonomic units. Nomenclature is concerned only with a taxonomy—one that can be used to identify the group to which each individual population, species, or higher taxon belongs. The different methods used to describe and diagnose a trilobite and a fish do not preclude the fish and trilobite from being classified as animals. In biogeography, the same procedure may be used to describe and diagnose "place" as an endemic area—that is, as a unit of biogeographical classification.

To treat endemic areas as units of classification, we must establish an area taxonomy that allows us to communicate our concept of an area to other biogeographers. We may call the smallest unit in our classification a district, which is part of a region, which in turn is part of a realm. We define an area within a classification (see Chapter 2).

The first global area classification of Candolle (1820) divided the world into 20 botanical regions. Almost 40 years later, Sclater (1858) divided the world into six regions, largely based on the distribution of birds. Wallace (1876), Merriam (1892, 1899), and others understood that the definitions of areas and regions based on the distributions of one kind of organism could be applied to all organisms. Although this was a forward-looking idea, it was fraught with practical problems, as we reviewed in Chapter 2. Merriam did not support larger regions and, instead, endorsed a non-hierarchical series of smaller, variably sized *life-zones*, a classification philosophy still used today (see Chapter 10). Although attempts were made to standardize "place" based on a classification, many scientists did

Box 3.1 International Code of Area Nomenclature (ICAN)

In 2007, the Systematic and Evolutionary Biogeographical Association (SEBA) drafted and adopted the first International Code of Area Nomenclature (ICAN). The code is based loosely on the International Code of Zoological Nomenclature. Its purpose is to stabilize the names of areas in biogeographic analyses (Viloria, 2004:164, 2005; Ebach et al., 2008). The aim is to have each name represent a biotic area that is accompanied by a diagnosis and description. The ICAN does not dictate how a classification is built (artificial versus natural or monophyletic) nor does it give preference to any particular method (e.g., species concept) used to diagnose taxa, and hence areas. The latest version of the ICAN is available at www.seba.uac.pt.

not agree on what they were classifying. Was "place" simply distributions based on historical events (such as geological processes and speciation), as Candolle believed, or was "place" based on the ecology and geography of an area, as Merriam believed?

Endemic areas are defined by their inorganic and organic elements. Once a taxonomic area unit is established, it may be compared to other areas. Most important, this also standardizes area names. A simple example demonstrates the importance of standards. "Borneo," the name of the world's third-largest island, has been used in many comparative biogeographic analyses: the "Borneo" of one study is compared to the "Borneo" of another. One problem with using "Borneo" in a comparative biogeographic analysis is that it is a named *geographic,* not organic, area. This is not unique to comparative biogeography as, for the sake of communication, biogeographic analyses begin with readily definable or recognizable areas. Because "Borneo" has not been standardized under a biogeographical area taxonomy, it means different things to different people. Consider biogeographic analyses that use "Borneo" as an area. Different areas are being compared: Parenti (1991) recognized "northwestern Borneo" as an area separate from the rest of the island; Andersen (1991) treated "Borneo" as part of the Malayan area that includes Sumatra, Java, Thailand, and the eastern coast of India; Vane-Wright (1991) and Michaux (1996) both considered "Borneo" its own, separate area, whereas Hennig (1966) combined "Java and Borneo" into one area. Were we to use the "Borneo" of each of these authors in a systematic biogeographical analysis, we would not be comparing the

Figure 3.4. The Australian turpentine tree, *Syncarpia glomulifera*, Albert Parade, Ashfield, Sydney, New South Wales. [Photograph by Giesela Dohrmann.]

same area, but sets of different areas that do not even necessarily overlap. A critical biogeographical comparison of "Borneo" to other areas of the world is impossible until there is an agreement on what areas are being compared (see Chapter 9).

DISCOVERING BIOTIC AREAS

Diagnosing Endemic Areas

Any biogeographical analysis that focuses on a broad region—say, the Pacific Ocean—must delimit a series of endemic areas. An ocean, like any habitat, is difficult to define based on organic distributions alone. Most taxic distributions overlap or include one or more smaller

Box 3.2 And Then There Was One...

Sturgeons and paddlefishes (the order Acipenseriformes) are among the most primitive actinopteryigan (ray-finned) fishes. Paddlefishes (family Polyodontidae), endemic to North America and China, have a rich fossil record dating from at least the Early Cretaceous of China (Grande et al., 2002). The two extant species, the paddlefish *Polyodon spathula*, Mississippi River Basin, North America, and the Chinese paddlefish, *Psephurus gladius*, Yangtze River and environs, China, were classified with fossil paddlefishes by Grande and Bemis (1991). These two species, although technically each other's closest living relative, were not hypothesized to be sister taxa; fossil paddlefish from western North America (Montana and Wyoming) were more closely related to *P. spathula* than to *Ps. gladius*.

Polyodon spathula is now likely the only living paddlefish, as numbers of the critically endangered *Psephurus gladius* have dwindled. Extinction can carve away portions of the living range of polyodontids, but can never alter their area of endemism.

distributions. Biotic distributions are also dynamic and rarely adhere to strict, recognizable boundaries; they expand and contract over time. A distribution may also split into smaller, localized units—say, populations—that later are reunited. Inorganic areas—the geography (including topology), climate, and geology (soil types and chemistry)—affect taxon distributions. Together, organisms and their environment form organic areas.

Organic areas are dynamic and change through the course of their history. The turpentine tree (*Syncarpia glomulifera;* Figure 3.4) grows only on dense soils in eastern Australia. Some trees grow out of the limestone rock itself. The weathering and erosion of limestone forms dense, clay-like soils on which many turpentine trees live. Erosion may, in time, weather underlying rock to form a substrate that becomes unfavorable to turpentines, such as coarse sandstones that erode to sandy soils. The relationship between the turpentine and its inorganic environment demonstrates Croizat's dictum (Chapter 1) that Earth and its life evolve together.

As the inorganic characteristics of an area change—as areas weather and soil types are modified—so does the biota that live in the area. Adjacent taxa that can live in the new environment may expand their range into the area, and the resident taxa may retract

Box 3.3 There's No Place Like Home

Endemism was key to the heralded discovery of *Tiktaalik roseae*, a fossil lobe-finned fish interpreted as a form intermediate between fish and amphibian (Daeschler et al., 2006). Well-preserved Devonian fossils some 375 million years old were discovered in 2004 on Ellsemere Island, Canada, inside the Arctic Circle. Of all the places in the world, how did scientists looking for fish–amphibian intermediates know to look there?

Ideal conditions for transitional fossil forms—age of the rocks (Devonian), type of rocks (sedimentary, as in a river bed), and exposure—come together notably in the sites of three former river deltas in North America: the Catskill Formation, East Greenland, and the Arctic Islands (Shubin, 2008). The first two of these were well studied; the Arctic Islands site was not. Predicting that *Tiktaalik roseae*, or something like it, would be found on Ellesmere Island illustrates the power of *place* in evolutionary biology.

their range and, in so doing, leave the area. This type of succession is common along coastlines or larger river deltas. Coasts recede and transgress, and rivers wander, changing their positions throughout time. The gradual or, in cases of extreme environmental change (such as on volcanic islands), relatively quick changes may result in overlapping distributions separated in time exhibiting *temporal displacement*. An example of temporal displacement between two areas is the limestone that forms the soil on which the turpentine grows and the trilobite that is fossilized in the rock. The trilobite and the turpentine both inhabit or inhabited different environments that share the same geographical space in Australia, separated by time.

The delimitation of an endemic area may be based on inferred geographical barriers such as a plateau, mountain chain, or river basin. The limits of a particular environment, such as savannah woodland or the temperature, salinity, and depth of a marine environment, all contribute to its delineation, providing each endemic area with elements of a diagnosis and description. The characteristics that diagnose and delimit each area, by definition, differ. A woodland environment will be delineated by soil types and geographical limits such as lakes or mountains. A coral reef will be defined by water temperature, salinity, depth, and tides, among other parameters. No two areas are the same size or have the same defining characteristics, whether they are geographical, geological, climatic, or ecological.

From the earliest considerations of biological distributions, it has been recognized that defining areas based solely on ecological parameters—such as amount of rainfall, density of groundcover, salinity, temperature, and so on—results in areas with different histories being defined in the same way. "Buffon's Law" states that different areas maintain different species despite similar ecological parameters. No two cool, temperate, wet sclerophyll forests are exactly alike in their taxon composition despite their similar ecologies. Organisms that are not closely related may look alike and may thus share analogous characters. The same is true for endemic areas and biota (see Chapter 4).

The *Prodrome* of Schilder and Schilder (1938–1939)

Franz and Maria Schilder were a team of mollusc systematists whose study of the family Cypraeidae, the cowries, led to an explicit statement on global areas of endemism. Paragons of a thorough, detailed systematic biology, the Schilders opened their 112-page paper on cypraeids with an apology that it contained "... the most important data only ..." (Schilder and Schilder, 1938:119), and therefore was titled a *Prodrome of a Monograph on Living Cypraeidae*. The *Prodrome,* as it came to be known, was a bold foray into marine biogeography unlike those that had come before. The Schilders identified geographically distinct races (or subspecies) and recognized them taxonomically (Schilder and Schilder, 1938:120):

> Such divisions of species into geographical races are familiar to students of terrestrial and freshwater animals, but their methods rarely have been used before in monographs on marine animals, probably on account of the wrong suggestion that the relative absence of natural barriers in distribution might prevent the development of geographical races.

They tied the name of a taxon to the area in which it lives: "No variety ... which is not a geographical one, is worth naming" (Schilder and Schilder, 1938:121).

Schilder and Schilder's global areas of endemism, as recognized on the distribution of endemic cowry taxa, contrasts with Sclater's biogeographic regions (Chapter 2) notably in the division of continents and large islands: Australia is not one area, it is three. Borneo, Sulawesi, and New Guinea are not whole; each is divided into two areas. Their description of the biology and distribution of cowries worldwide could thus be

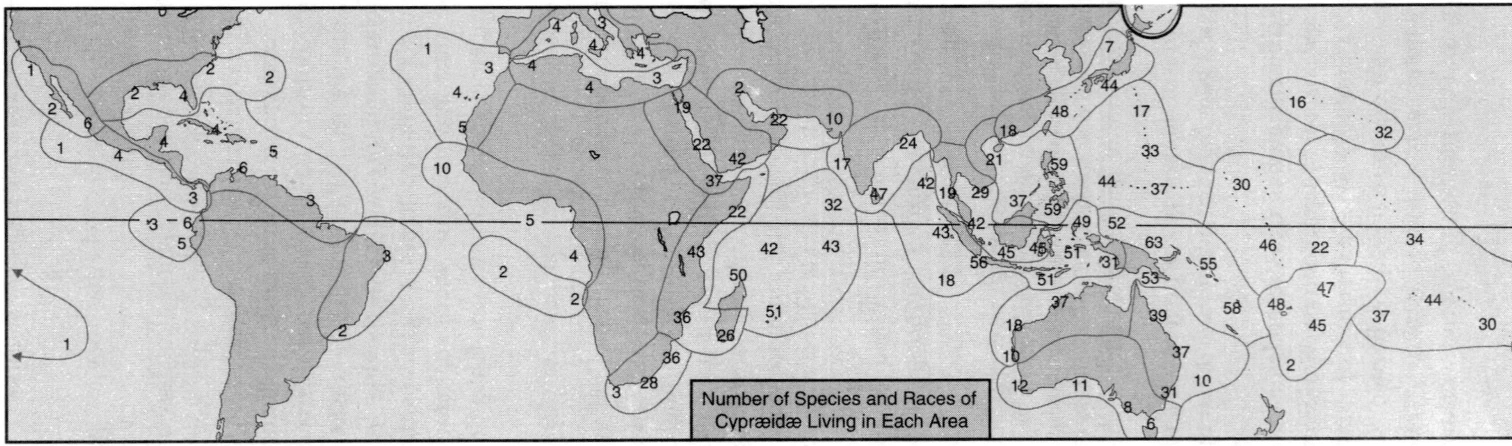

Figure 3.5. Thirty-one global biogeographic regions, as exemplified by numbers of species and races of marine molluscs of the family Cypraeidae, the cowries, following Schilder and Schilder (1938:223, Map 2). [Modified map reproduced with permission of Oxford University Press and the Malacological Society of London.]

Box 3.4 Franz Alfred Schilder (1896–1970) and Maria Hertrich Schilder (1898–1975)

Figure 3.6. Franz and Maria Schilder. [Photographs from *Hawaiian Shell News* (1964:3), courtesy of Wesley Thorsson.]

Franz and Maria Schilder were a dedicated and prolific team of mollusc systematists. Franz was an Austrian who trained at the University of Vienna where he began his unrivaled collection of cowry shells. After emigrating to Germany in 1922, he married Maria Heitrich, a German chemist who switched her professional focus to the study of molluscs. Franz Schilder held several academic appointments in Germany, notably at the University of Halle/Saale, and also taught at the University of Leipzig. Together, they produced over 250 scientific papers, most on the living and fossil Cypraeidae, or cowries. Franz Schilder (1956) published *Lehrbuch der Allegemeinen Zoogeographie*, a text which focuses on the recognition of biogeographic areas at all geographic scales.

used as a prodrome, a precursor or outline, of global areas of endemism (Figure 3.5). The areas are significant because they are organic areas, defined not just by geography, but by the geographic, and geologic, limits of taxa. We test Schilder and Schilder's areas of endemism hypothesis for the Indo-Pacific in Chapter 9 to implement the principles of a comparative biogeography.

Testing Proposals of Endemic Areas

As taxa initially may be diagnosed on analogies, rather than homologies, so may areas. Once an area of endemism is proposed, its relationship to other areas may demonstrate that it is not "monophyletic." This means that an area has one set of relationships to some areas, and another set of relationships to other areas. Real world examples of such area relationships may be extremely complex; they seemingly defy explanation and elude pattern. In Chapter 4, we address some of these complex relationships with worked examples. We demonstrate how discovery of biotic area homology, which for three areas A, B, C might be expressed as C(AB), is key to analyzing the amount and kind of information we have about endemic areas, their definition, and their relationships.

SUMMARY

- Endemism is a key concept in biogeography that specifies the relationship between organism and place.
- An endemic area is any disjunct or continuous geographical space, through time, that delimits the current and past distribution of one or more taxa.
- The qualitative (e.g., relationship) rather than the quantitative (e.g., similarity) aspects of biotic areas are critical for defining endemic areas.
- Area nomenclature is vital in communicating about areas. Because ways of defining areas vary, the naming of areas requires a single system or set of rules. A nomenclature makes effective communication possible.

NOTES

1. Prior to a theory of biological evolution, an area of endemism was considered to be the area in which an organism was created (Kinch, 1980).

2. The ecology of invasive species, especially the interaction between native and introduced taxa, is becoming a critical avenue of investigation, especially when introductions decimate the native biota (see, e.g., Sax et al., 2005).

3. The refuge theory was proposed initially as a mechanism of speciation of South American birds: alternating wet and dry cycles during the Pleistocene restricted birds to small, wet regions or refuges, where they lived and differ-

entiated until the wet regions expanded and the birds likewise expanded their ranges (see Haffer, 1969; Prance, 1982). The concept has been generalized in biogeography to refer to the survival of a taxon in a restricted area through a geologically, including climatologically, adverse period.

4. Three late-19th- to early-20th-century German biogeographers, Adolf Engler, Georg Oskar Drude, and Ludwig Diels, defined a Pacific Austral flora, characterized by terrestrial taxa such as the southern beeches (Nothofagaceae) and by other genera now scattered over a vast area spanning the southern oceans. Andrés Moreira-Muñoz (2007) examined Engler's (1882) classification of the Austral Realm, which excludes marine flora and fauna, revealing the problems that a fragmented flora may pose to a biogeographic classification: do the fragments represent relicts of a once more widely distributed biota or do they represent individual episodes of colonization?

To simplify Engler's (1882) classification, Cox (2001) refined the biogeographic areas by restricting terrestrial realms to continents, rather than having them span oceans. In so doing, Cox dismissed transoceanic biogeographical patterns and resurrected pre-tectonic notions of long-distance dispersal pathways to explain the distribution of a fragmented biota (see Axelrod, 1972; Moreira-Muñoz, 2007). Long-distance dispersal, like all other mechanisms, should not be part of an area classification. Following in the tradition of marine zoogeographers of the 19th century (see Chapter 2), Morrone (2002) recognized three global biotic realms—Holoarctic (Boreal), Austral, and Holotropical (Pantropical)—thus salvaging Engler's (1882) notion of a Pacific Austral region.

5. Global patterns of species diversity have traditionally been explained from either an ecological or an historical viewpoint, each of which has often been considered contradictory or irrelevant to the other. Solely ecologically based models, such as those centered on contemporary climatic factors, were found to be inadequate for predicting diversity patterns of endemic birds of South America by Rahbek et al. (2007). They concluded that such models underestimate the importance of historical factors in determining continental species diversity.

6. The terms *native, endemic, indigenous, aboriginal,* and *autochthonous* are often treated as synonyms, although they are not strictly so. Indigenous, for example, means that an organism is native to an area but that it may also live elsewhere. Autochthonous refers to an organism having originated where it is found. In geology, autochthonous rocks are those formed where they are found (see Chapter 8).

7. International Code of Zoological Nomenclature, Fourth Edition, 1999; International Code of Botanical Nomenclature, the Vienna Code, 2006; International Code of Nomenclature of Prokaryotes.

FURTHER READING

Cecca, F. 2002. *Palaeobiogeography of marine fossil invertebrates. Concepts and methods.* Taylor & Francis, New York.

Everhart, M. J. 2005. *Oceans of Kansas. A natural history of the western interior sea.* Indiana University Press, Bloomington and Indianapolis.

Longhurst, A. R. 1998. *Ecological geography of the sea.* Academic Press, New York.

Patterson, C. 1983. Aims and methods in biogeography. In R. W. Sims, J. H. Price, and P. E. S. Whalley (eds.), *Evolution, time and space: The emergence of the biosphere* (pp. 1–28). Systematics Association Special volume 23, Academic Press, New York.

FOUR

BUILDING BLOCKS OF BIOGEOGRAPHY

Biotic Areas and Area Homology

BIOTIC AREAS AND AREA HOMOLOGY

The role of biotic areas in biogeography is analogous to that of taxa in systematics. Taxa are defined by homologous relationships of their organic parts, called homologs. Similarly, biotic areas are defined by the homologous relationships of their endemic areas. These biotic components, or parts, are area homologs, analogous to taxa and their relationships (e.g.,

Overview

Relationships among biotic areas are recognized as area homologs and area monophyly. The smallest item of relationship among areas, a three-item relationship, is an area homolog. Corroboration of area homologs is area homology. Area homologs that make up larger area relationships may overlap to form many area homologies. The combination of overlapping area homologies is area monophyly (geographical congruence): a natural biogeographic classification.

Finding area homologies and monophyly is complex in practice. Area analogies are common, as are Multiple Areas on a Single Terminal-branch (MASTs) and areas present in one clade and missing from another. Temporal overlap, when what is initially recognized as a single biotic area is found to contain two or more unrelated biotic areas, confounds the search for area homology.

The structure of an areagram—that is, the hierarchy of areas—plays a crucial role in determining whether the areagram is informative or not. Areagrams with unresolved crowns (deeply nested components) are uninformative, whereas areagrams with unresolved basal components are still partially informative.

Information can be extracted from areagrams to propose hypotheses of area relationship that can be tested. Geographic paralogy, the repetition of areas on an areagram, provides no information on area relationships.

Morrone, 2001a). Relationships among biotic areas constitute the natural classification of place that can be used to test whether we have defined our biogeographic units, such as areas of endemism, in an informative, meaningful, and natural way. Area classifications, therefore, are derived directly from systematic classification. Most important, the biotic area characters *are* taxic relationships, which makes the study of biotic areas dependent on the discovery of natural, monophyletic groups of organisms.

Definition of a Biotic Area

A biotic area consists of homologous area relationships expressed by more than one monophyletic group that inhabits a common place and/or a designated endemic area. Biotic areas are part of a natural or monophyletic classification. They do not represent any particular evolutionary, geographical, biological, or geological mechanism of distribution, but may be used to infer a mechanism or to choose among possible mechanisms.

Definition of Area Homology

An area homolog is the smallest unit of meaningful cladistic relationship among areas. If just two areas are compared, their taxa might be closely related or distantly related; there is no way to decide. If two areas, A and B, each have a species that is more closely related to the other than either is to a third species living in area C, then areas A and B are hypothesized to share a history that is not shared by area C. The three-area relationship, or area homolog, is C(AB).

DISCOVERING BIOTIC AREA RELATIONSHIPS

To discover biotic area relationships, we need to establish a description and classification of endemic areas to be studied. The naming of areas may be governed by an area nomenclature that minimizes ambiguities, such as the International Code of Area Nomenclature, or ICAN (see Chapter 3). Our goal here is not to propose or outline a comprehensive global area nomenclature.[1] We instead outline a step-by-step procedure for discovering biotic areas.

Describing Area Homologies and Discovering Area Monophyly

An endemic area must first be diagnosed and defined. With an area nomenclature and classification, an endemic area may be compared to others. Because endemic areas are described in part by their inorganic characteristics, we require the feature that all areas share: taxic relationships. By comparing endemic areas, we can test for area homology (see Chapter 2). Area homology is found by comparing area homologs among endemic areas to discover geographic congruence. We illustrate these principles with several real examples.

Species of the marine water-strider genus *Halobates* (family Gerridae; Figure 4.1a) are distributed broadly throughout the Indo-West Pacific, as illustrated for the *Halobates regalis* species group (Figure 4.1b; Andersen, 1991). The areagram for this species group is complex, with overlapping, redundant, or geographically paralogous areas in Australia. The species that lives in the Philippines and Solomon Islands (species area 5) is related more closely to a subgroup of species in northern Australia (species areas 6, 7, 8) than it is to a water-strider from Malaysia and Sri Lanka (species area 4). The area relationship Malaysia/Sri Lanka (Philippines/Solomons, northern Australia) or species areas

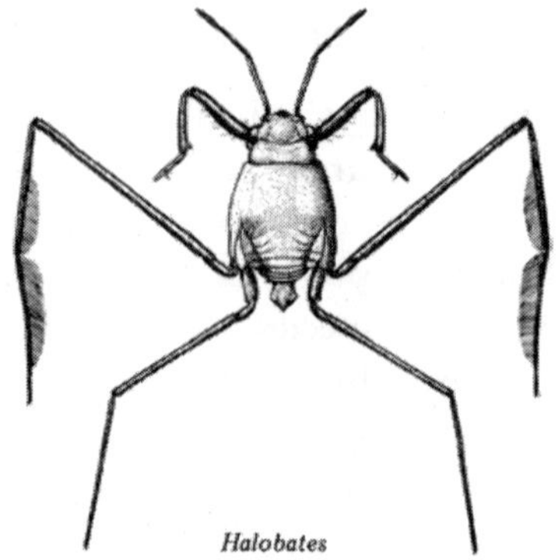

Figure 4.1a. ***Halobates micans*, a marine water-strider (from Andersen, 1991: Fig. 1). [Reproduced with permission of CSIRO Publishing.]**

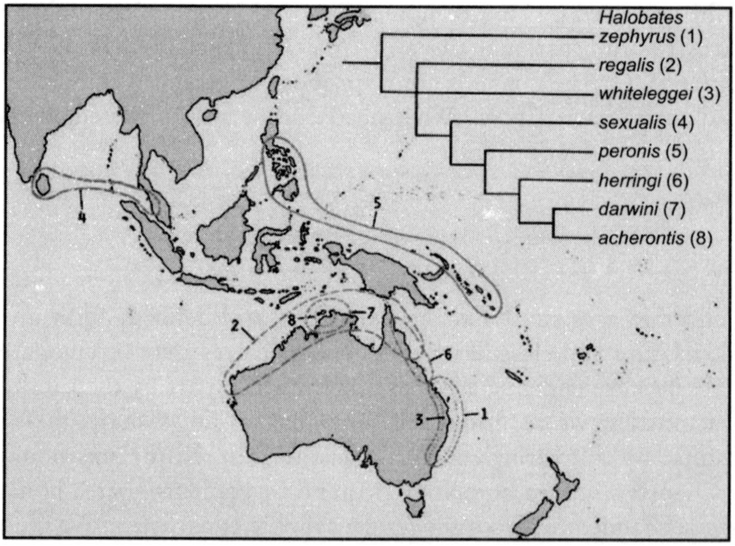

Figure 4.1b. Distributional ranges (outlines numbered 1–8), cladogram, and areagram of the eight species of the *Halobates regalis* group of water-striders (modified from Andersen, 1991: Fig. 7). [Reproduced with permission of CSIRO Publishing.]

4(5(6(7,8))) is derived from this taxic relationship. The distribution and relationships of species areas 6(7,8) form an area homolog for regions of northern Australia. An area homolog may be corroborated if it is found in other areagrams derived from other monophyletic groups that live in the same regions. An area homolog is analogous to a character homolog in systematics; area relationships are hierarchical, as are character and taxon relationships.

Area homology, or area monophyly, is discovered when different areagrams are compared and their area homologs corroborate those of other areagrams. The resulting statement of area relationship that forms a natural classification of areas is a *general areagram.*

Using Area Monophyly to Test Existing Classifications

The discovery of area monophyly corroborates homology of the areas of endemism used in the analysis. Unresolved nodes in the general areagram (not present in the original areagrams) indicate that some areas or groups of endemic areas are defined poorly: they are analogs, not homologs. General areagrams are analogous to consensus trees of multiple, single-character cladograms, or gene trees in systematics. Consensus cladograms in systematics can indicate for which taxa there is evidence of monophyly or not. This is also true for biotic areas in general areagrams. Complexities of areagrams, such as area analogy, paraphyly, unresolved nodes, and conflicting patterns, are best resolved either by redefining the endemic areas or by restating the question being asked.

THE REAL WORLD: COMPLEXITY OF AREAGRAMS

Areagrams represent the complex relationships among biotic areas through time. Practically, they exhibit various degrees of resolution, in part because of differential responses of taxa to Earth history events, to extinction of lineages, or to the failure to collect all taxa. An areagram may not have all nodes resolved, may have areas duplicated, or may contain areas with taxa that live outside the area of interest. More frustrating in practice is the sparse number of areagrams with enough areas in common to propose area homologies. In each case, the comparative biogeographer should assess the *quality* of the data at hand, as below.

Area Analogy

Systematists scrutinize what data go into and what stay out of their analyses. The same should hold for the data (taxa and areas) used in a biogeographical analysis. A character in a systematic analysis that relates all taxa *equally* is uninformative and is therefore dismissed. The same is true for area relationships, and notably so for a two-area statement of "relationship." Any two areas, or indeed objects, may be

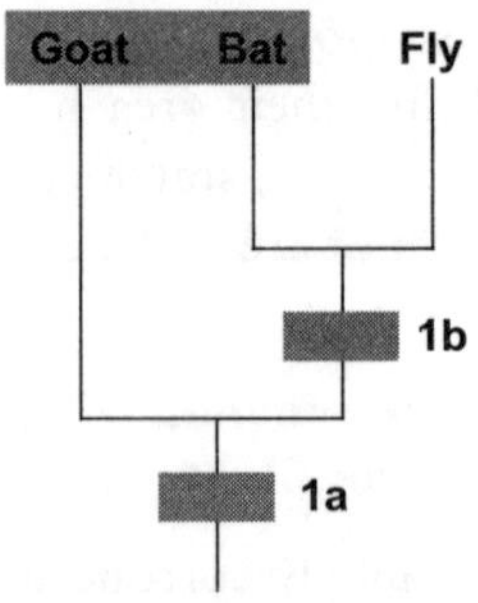

Figure 4.2. Example of an analogous node (1b). A goat, a bat, and a fly all have anterior appendages, character state 1a. Based on other data, the mammals (green)—goat and bat—are considered to be more closely related to each other than either is to a fly, despite the character state that the bat and the fly share (1b, wings present).

"related," but their relationship is not informative. It is a measurement of similarity. Two-area statements of relationship in biogeography are uninformative as are two-taxon statements in systematic analysis. Only when two areas are compared to a third can we recognize an informative area relationship.

Analogous characters are common in systematics. An analogous character (analog) is one that *appears* to be homologous but is not (Figure 4.2). It relates two taxa together, when compared to a third, based on a false assumption of homology. The presumptive homolog may be described poorly and, by definition, is present in unrelated taxa. Biogeography has "area analogs." An area analog is singular in its statement of relationship; the presumptive homolog is found in no other areagrams. Area analogs specify unique relationships of areas and, like their systematic counterparts, do not form patterns and therefore cannot be used to discover area homology. Area homologs can also conflict due to poorly defined endemic areas, rather than area analogy, or they can reflect area relationships from different time periods (see below).

MASTs (Multiple Areas on a Single Terminal-branch) and Geographic Paralogy

A taxon that lives in more than one area of endemism is called widespread. If we assume that dispersal is possible, we cannot know whether the taxon is native to all of the areas of endemism combined or to a subset of those areas. Widespread taxa may share phylogenetic relationships with taxa in the same or different areas. To discover these area relationships, we take a systematic approach. If we take any other approach, such as modeling for endemicity or searching for the "true" area using

Figure 4.3a. The Indo-Pacific scleractinian coral, *Acropora selago*, Ribbon Reef 9, Great Barrier Reef, December 9, 2007. [Photograph by Paul Muir.]

some arbitrary set of criteria, such as the largest area or the area closest to a presumed source, then we abandon a comparative method to resolve area relationship. Instead, we propose to resolve widespread taxa in a comparative biogeography by *general comparison*.

The objective of systematic biogeography is to discover a hierarchical set of area relationships expressed in a classification. Systematic biogeography is an historical science. It is impossible to *know* what distributional events have taken place; they must be presented as hypotheses. As systematic biogeographers, we are obliged to generalize—that is, to form a classification based on relationship, not to propose an untestable scenario that satisfies our need for an explanation but that ignores or contradicts area relationships.

Acropora (Figure 4.3a), the most diverse reef-building coral genus, with over 100 extant species worldwide, is particularly abundant in the central Indo-Pacific. Comparison of the relationships and distribution of species and species groups of *Acropora* throughout the Indo-Pacific, in particular *Acropora selago* (Figure 4.3b), along with the studies of *Halobates* (above), were among the first applications of cladistic biogeographic methods to marine invertebrates (e.g., Wallace et al., 1991).

Four somewhat overlapping biogeographical areas were identified to describe the distribution of five species within the *A. selago* group

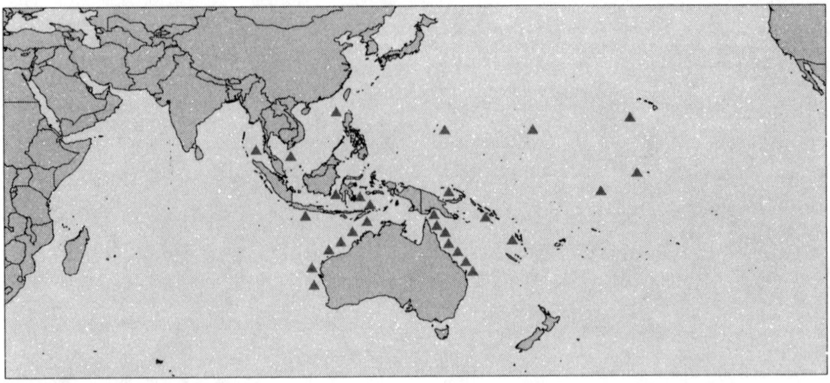

Figure 4.3b. Distribution of the Indo-Pacific scleractinian coral, *Acropora selago* (redrawn from Wallace et al., 1991: Fig. 5, top map).

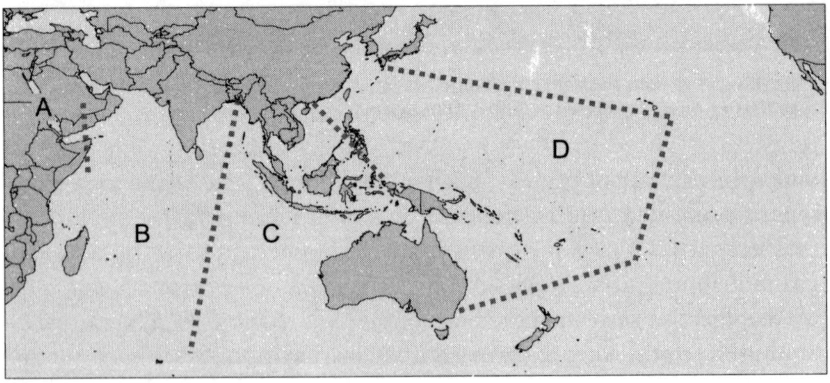

Figure 4.4. Biogeographical areas of species of the *Acropora selago* species group (redrawn from Wallace et al., 1991: Fig. 6). A = Red Sea; B = western to central Indian Ocean; C = eastern Indian Ocean; D = western to central Pacific Ocean.

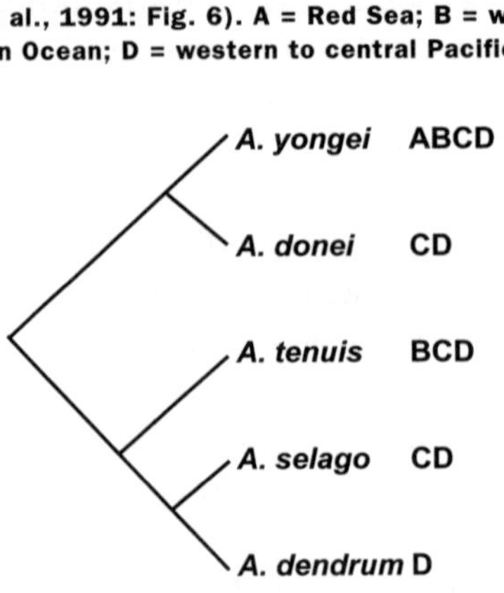

Figure 4.5. The cladogram and areagram of five species of *Acropora* (redrawn from Wallace et al., 1991: Fig. 4). Areas A, B, C, and D are as in Figure 4.4.

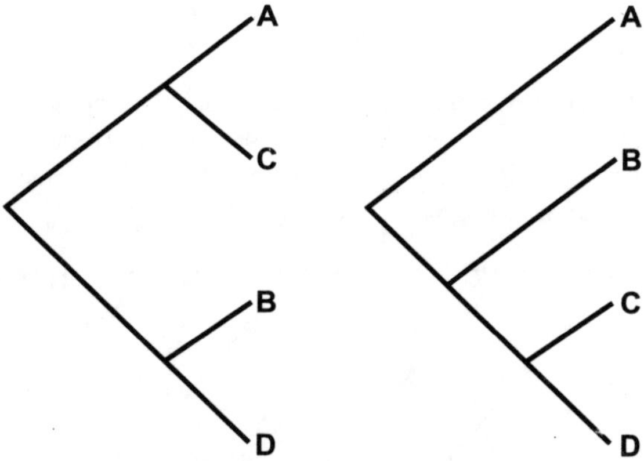

Figure 4.6. Two possible resolved areagrams for areas A, B, C, and D (as in Figure 4.4) derived from the areagram of Figure 4.5 (redrawn from Wallace et al., 1991: Fig. 7).

(Figure 4.4): Red Sea; western to central Indian Ocean; eastern Indian Ocean; and western to central Pacific Ocean. All but one species, *A. dendrum,* is widespread (lives in more than one area) in the areagram for the five species in the *A. selago* group (Figure 4.5). Relationships of widespread taxa are represented diagrammatically using parentheses to separate or to group taxa or areas and using an underscore for MASTs (multiple areas on a single terminal-branch): the areagram of the species group is represented as (ABCD, CD)(BCD(CD, D)).

We extract statements of area homology using the relationships among the areas as specified in the areagram. Information on the relationships of area A comes from just one taxon, the widespread species *A. yongei,* that lives in all four areas. To extract the information on area relationships, we treat each area as a separate occurrence. Among the possible areagrams that summarize the information given in the relationships represented as (ABCD, CD)(BCD(CD, D)) are two different areagrams: (AC)(B(CD)) and (AD)(B(CD)). They share one area homolog: B(CD). As for the relationship of A, there is insufficient information to choose among several alternatives, such as A(B(CD)) or (AC)(BD), as concluded by Wallace et al. (1991; Figure 4.6).

Overlap of closely related taxa can also introduce ambiguity into a biogeographic analysis. Closely related taxa that live in one area are *geographically paralogous*. Geographic paralogy is uninformative. There are many reasons for geographic paralogy: extinction, failure to collect, too broad or too narrow area delineation, and so on. In evaluating an area classification, these reasons should remain separate from the discovery

Box 4.1 Geographic Paralogy

Geographic paralogy, represented by the duplication or overlap of areas related at a node, was identified by Nelson and Ladiges (1996) as one of the principal sources of error when extracting information on area relationships from areagrams.

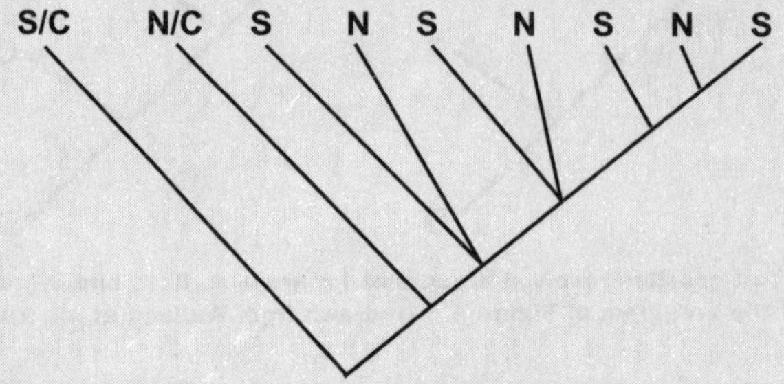

Figure 4.7. Areagram of a subgroup of cyprinodontiform fishes (following Parenti, 1981: Fig. 93).

This areagram is of aplocheiloid killifishes living in Central and South America (northern and southern regions of South America distinct across a range of taxa; Parenti, 1981: Figure 93) is rife with geographic paralogy. The three areas of endemism are Central America (C), northern South America (N), and southern South America (S). All three areas demonstrate geographic paralogy. The taxon widespread in C and N is sister to a widespread lineage in N and S. When compared with the widespread taxon in C and S, it provides one informative statement on area relationships among the areas, or one area homolog: S(NC).

Geographic paralogy may have been caused by repeated extinction events in Central America, by lineage duplication, or by sympatry (dispersal). It is impossible to say which is the most likely explanation, or whether all three were involved, without comparing this area homolog with others from the same areas—in other words, without testing the generality of this area homolog. To choose among the explanations would be analogous to inferring the phylogenetic relationships among a group of taxa from one character-state tree.

and analysis of pattern. Removing geographic paralogy requires refining hypotheses of relationships or redefining areas of endemism.

The tropical and subtropical perciform fish family Cichlidae (Figure 4.8) and the pantropical and temperate atherinomorph fish

Figure 4.8. Threadfin acara, *Acarichthys heckelii,* family Cichlidae. [Photograph copyright John Brill.]

Figure 4.9. Green Swordtail, *Xiphophorus hellerii,* order Cyprinodontiformes, family Poeciliidae. [Photograph copyright John Brill.]

order Cyprinodontiformes (Figure 4.9) have largely coincident, overlapping distributions (Figure 4.10). Both clades figure prominently in biogeographic analyses because of their distributions and the intense scrutiny paid to their phylogenetic relationships during the past three decades (e.g., Stiassny, 1981, 1991; Farias et al., 2000; Parenti, 1981, 2005, and below).

The freshwater family Cichlidae is classified in four subfamilies (following the molecular phylogenetic hypothesis of Sparks and Smith, 2004) with these relationships: Etroplinae (Ptychochrominae (Cichlinae, Pseudocrenilabrinae)). The Etroplinae live in Madagascar and India, the Ptychochrominae in Madagascar, the Cichlinae in South America, and the Pseudocrenilabrinae in Africa (Figure 4.10a). The areagram for cichlids is:

India/Madagascar (Madagascar (South America, Africa))

Madagascar is geographically paralogous on the areagram. Geographic paralogy of Madagascar is uninformative because it does not specify a close relationship between India and Madagascar; it is simply redundant within the area classification. The areagram specifies one set of resolved area relationships: India (Madagascar (South America, Africa)). The informative three-area statements specified by this areagram are:

(South America, Africa) Madagascar

(Africa, Madagascar) India

(South America, Madagascar) India

Cyprinodontiformes are classified in two suborders, Aplocheiloidei and Cyprinodontoidei, which are largely sympatric, with both being broadly distributed in South America, Africa, and Madagascar (Figure 4.10b; Parenti, 1981). Aplocheiloids also live in South and Southeast Asia. Their phylogenetic relationships are compared to those of cichlids here in a search for a generalized distribution pattern, a test of biotic area homology.

Aplocheiloids comprise several hundred species of tropical freshwater fishes whose areagram, also based on molecules (Murphy and Collier, 1997) for the purpose of comparison with the cichlid analysis, is (Africa, South America), (Indo-Malaya, Madagascar/Seychelles). There is no geographic paralogy in the cyprinodontiform areagram: (Indo-Malaya, Madagascar/Seychelles) (South America, Africa). The four informative three-area statements specified by this areagram, with Madagascar and Seychelles combined for the purposes of discussion, are:

(Indo-Malaya, Madagascar) South America

(Indo-Malaya, Madagascar) Africa

(South America, Africa) Indo-Malaya

(South America, Africa) Madagascar

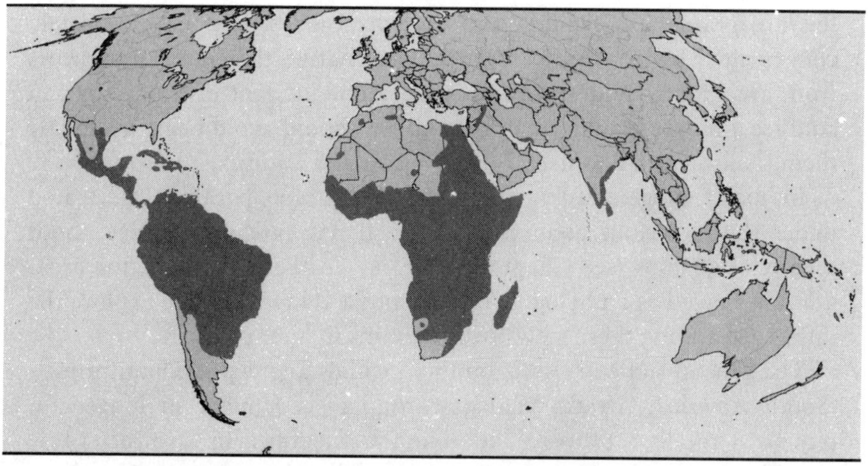

Figure 4.10a. Distributional limits (green) of the family Cichlidae (after Berra, 2001:440). [Modified map copyright Tim Berra.]

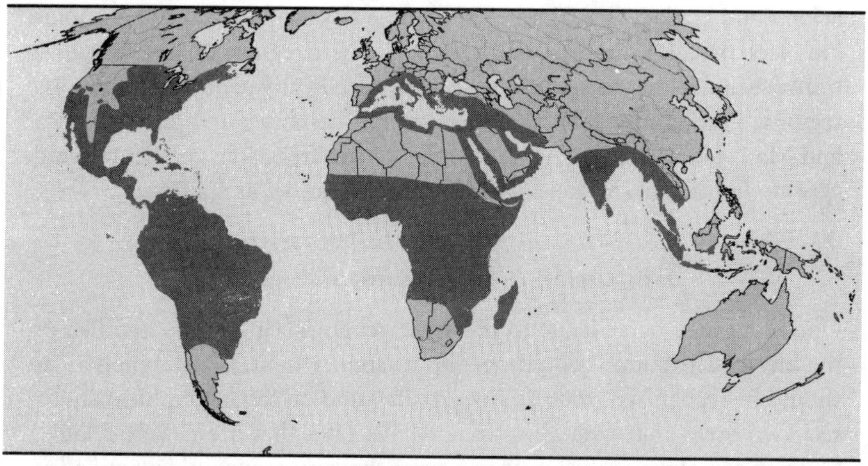

Figure 4.10b. Distributional limits (green) of the order Cyprinodontiformes (modified from Parenti, 1981: Fig. 1).

The last areagram is congruent with the first specified for the cichlids and, therefore, with a corroborated biotic area homology: (South America, Africa), Madagascar.

The areagrams based on molecular data for aplocheiloids and cichlids differ in amount of information they contain on the relationship of Madagascar. Again, Madagascar is geographically paralogous on the cichlid areagram. Had the cichlid subfamily Ptychochrominae gone extinct in Madagascar and had we never discovered their fossils, the areagrams for cichlids and aplocheiloids would be completely congruent. This is why we extract informative three-area statements from areagrams. Widespread, missing, or redundant distributions can confuse a biogeographic analysis, and we should avoid being misled by them (Nelson and Platnick, 1981; Nelson and Ladiges, 1996).

In many biogeographic methods, geographic paralogy is treated under the erroneous assumption that all data are informative about area relationships (see Chapters 5 and 6). In these methods, the areagram is viewed as a phylogenetic tree, not a cladogram. We explore the difference among these views of areagrams in later chapters.

The general summary areagram for cichlids and cyprinodontiforms—(South America, Africa) Madagascar/India—is familiar in biogeography, as it recalls a trans-Atlantic biota as identified by Croizat (1958: also Figure 1.2 this volume). We could predict that further information would resolve the sister area relationships of Madagascar and India. Most important for a global classification, the classic biogeographic regions of Sclater and Wallace (see Chapter 2) occupied by cichlids and aplocheiloids are Neotropical, Ethiopian, and Oriental. These regions do not recognize transoceanic relationships, as they do not specify any relationships among regions. Furthermore, the Ethiopian region combines two areas, Africa and Madagascar, that are separate in all of the areagrams for cichlids and cyprinodontiforms. Sclater's classification is rejected as spurious.

Overlapping Areas: A Temporal Aspect?

One biota may be endemic to the same geographical area as another, or the biotic distributions could overlap to some extent. Overlapping areas should be recognized prior to any classification or biogeographical analysis. Two areas that overlap may not have taxa that are closely related. Overlap can cause confusion as two areas that are erroneously identified as one are *homoplastic*—that is, the same area is related equally to all others. The area homologs derived from homoplastic endemic areas are not themselves homoplastic, as they depict actual relationships that convey conflicting information about a particular endemic area. One extreme example of overlapping distributions is that of a living bird, the rockwarbler, *Origma solitaria,* and an extinct amphibian, the labyrinthodont *Paracyclotosaurus davidi* (Figure 4.11). Both taxa exist only on or within the Permo-Triassic (290–200 mya) Hawkesbury Sandstone of the Sydney Basin, Australia. Although both taxa share *the same geographical location,* their distributions are separated by over 200 million years. How do we incorporate this temporal discord into a biogeographical analysis?

Figure 4.11a. The rockwarbler, *Origma solitaria*, in its endemic habitat, sandstone formations, Curracurrong Falls, Coast Track, Royal National Park, Sydney, Australia. [Photograph by Caitlin Hulcup.]

Figure 4.11b. A reconstruction of the Triassic labyrinthodont *(Paracyclotosaurus davidi)* from the Sydney Basin. [Image copyright Rick Sardinha.]

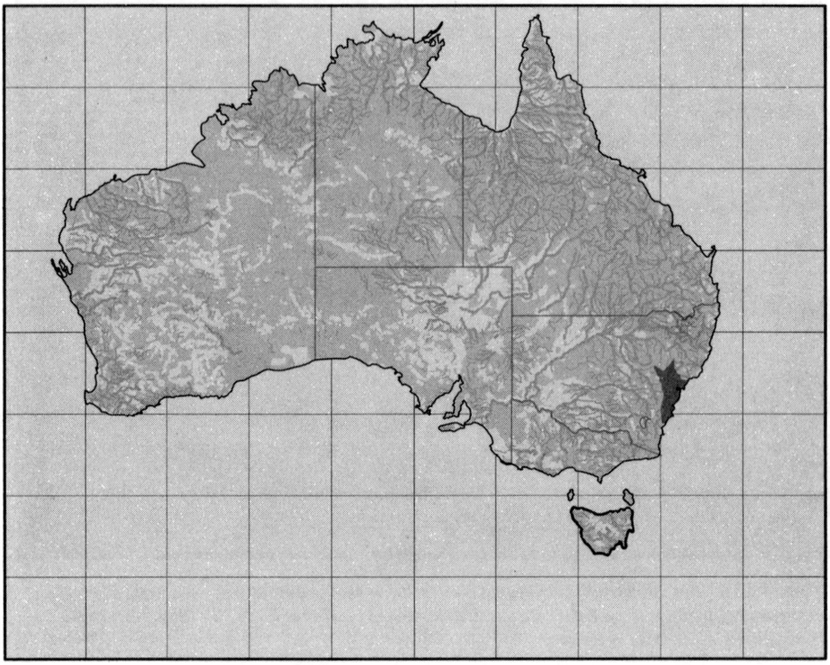

Figure 4.12. The geographical limits of the Sydney Basin (green).

The temporal aspect of biogeography has traditionally been split off into a separate field: paleobiogeography, the study of the age and distribution of fossils. More recently, relative timing of events has been considered superior to identification of general biogeographic patterns by many molecular systematists to explain biogeographic distributions (see Box 4.2). They compare sequence divergence times with hypothesized ages of geological events to test biogeographic hypotheses. There is no *a priori* justification for a temporal separation of distribution patterns, as all biotic areas can be classified in the same area taxonomy. Furthermore, such divisions add to the misunderstanding that without fossils or molecules biogeography lacks a temporal or geological aspect. One popular assumption is that most, if not all, modern taxa and their environments evolved relatively recently (from the Plio-Pleistocene to the present day) and have no significant history influenced by older geographical or geological processes. This assumption recalls the late 19th and early 20th centuries, when the Earth was still considered quite young, compared to our recent estimates, and when it was thought to be static, not mobile. Living clades have a varied history: some exhibit a

Box 4.2 Molecular Evidence of Divergence Times: Hypotheses versus Evidence

Correlation between molecular sequence divergence and divergence time (Zuckerkandl and Pauling, 1962) sparked a decades-long investigation into the use and application of molecular sequence data to estimate ages of lineages. Constant rate divergence, or a "molecular clock," has often been assumed, but is not always justified. Clock calibration remains controversial.

Molecular estimates of divergence times between and among taxa have been used extensively to accept or reject explanations for biogeographical patterns. Strong proponents of this method (e.g., de Queiroz, 2005) have argued that the relatively young estimates of the age of many lineages, especially those with trans-Pacific distributions, should be used as evidence to reject older, vicariant explanations in favor of more recent episodes of individual-clade dispersal. This argument has been countered by Heads (2005a) and others, who criticize some of the various ways divergence times have been estimated and the clock calibrated. As one example, calibration of molecular clocks using geological parameters, such as estimated ages of African rift lake basins for cichlid fishes (e.g., Vences et al., 2001), emphasizes their dependence on geology (see also Chakrabarty, 2004; Chapter 7) and makes subsequent comparison with some geological processes circular.

Equally important for biogeography, Heads and others have reiterated that estimated divergence times are *hypotheses,* not empirical observations: they cannot provide *evidence* for long-distance dispersal. Estimates of the timing of plate tectonic and other geological processes are themselves *hypotheses,* not empirical observations, and are likewise subject to discussion, debate, and refutation (e.g., McCarthy, 2003, 2005; McCarthy et al., 2007).

distribution pattern that is coincident with geographical limits, whereas others have distributional ranges that may conform to inferred barriers that are no longer identifiable on a modern landscape. *Acropora* fossils from the Paleocene of Somalia (western Indian Ocean) and mid-Eocene of the Mediterranean and Atlantic/Caribbean demonstrate the diversity and broad distribution of the genus long before the Plio-Pleistocene (Wallace and Rosen, 2006).

To return to the cichlids and cyprinodontiforms, both molecular analyses (Sparks and Smith, 2004; Murphy and Collier, 1997) concluded that each lineage was old enough to have been affected by geological events that are part of the traditionally accepted sequence of the break-up of the supercontinent Pangea (Rosen, 1974: Figure 44) and that, therefore, both distribution patterns were consistent with the mechanism of vicariance. Alternatively, another analysis of molecular sequence data (Vences et al., 2001) concluded that cichlids were too young to have been affected by the Pangean break-up and that portions of the distribution pattern should therefore be explained by independent, long-distance dispersal events. Details of the phylogenetic biogeographic patterns are not in dispute—just the timing of differentiation. This discord has been referred to as *pseudocongruence* (Donoghue and Moore, 2003): *congruence* because the patterns are the same, *pseudo* because the inferred cause (because of the inferred age difference) is different. Some argue that biogeographic analyses should therefore segregate taxa by age. One problem with such an *a priori* sorting of taxa by age—for example, when trying to construct an analysis of Eocene fish biogeography—is that estimated divergence times for taxa are highly variable. The living coelacanths, genus *Latimeria,* for example, are represented by two disjunct species, one in the western Indian Ocean, the other in the western Pacific, with estimates of divergence times ranging from 1.3 mya to 40 mya (see Parenti, 2006). Furthermore, and even more important for a comparative biogeography, phylogenetic relationships, and therefore area relationships, are hierarchical, as demonstrated above for the water-striders *Halobates*. Coelacanths, as a lineage, date from at least the Upper Devonian (Forey, 1998). Was the distribution of the two living species affected by recent events, ancient events, or a combination of the two? Because we cannot answer that question at the outset, we are obliged to consider all distribution data and taxonomic relationships to discover biogeographic patterns.

Introducing Ambiguity: Polytomies in Areagrams

One goal of comparative biogeography is to minimize ambiguity. Ambiguous area relationships add nothing to a biogeographic analysis and may give erroneous results. Ambiguous data, such as those contained in polytomies, cannot be resolved using comparative methods. Instead, they generate uninformative data of erroneous area relationships. Every trichotomy includes two relationships for which there is no evidence; polytomies among four or more terminals generate even more possible relationships.

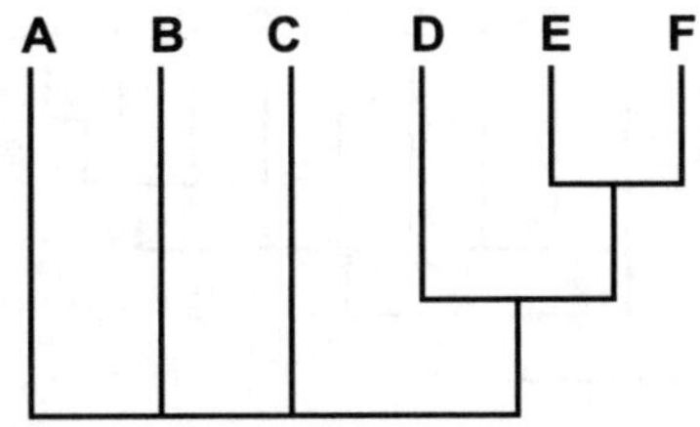

Figure 4.13. An example of a basal polytomy in an areagram. The single, informative subtree is D(EF).

Area relationships can be either informative or simply redundant—conveying data that do not add to or take away from an area homolog. Areagrams plagued with geographic paralogy and MASTs may be resolved by extracting informative area relationships. Polytomies in areagrams should be pruned prior to resolving area relationships in geographical paralogous nodes or MASTs. In any areagram, there are three possible positions for a polytomy: basal, nested, or crown. Below, we give hypothetical examples of the type of polytomies found within an areagram. Following the hypothetical examples, we discuss real areagrams that demonstrate some of these different types of polytomies.

Basal Polytomies

To resolve areagrams with basal polytomies, one must prune the basal node that contains the ambiguity. In Figure 4.13, for example, the basal polytomy ABC contains ambiguous information about the relationships of areas A, B, and C to each other and to areas D, E, and F. There is no way to decide among the following possible area resolutions: A(D(EF)), B(D(EF)), or C(D(EF)). Each is equally likely. There is just one resolved, informative subtree in this areagram: D(EF), an area homolog.

Nested Polytomies

Nested polytomies are resolved through a method similar to that used for basal polytomies. In the hypothetical example in Figure 4.14, there is a nested polytomy, CDEFG, and one resolved, informative subtree: H(IJ). Including the basal areas in the resolved areagram, CDEFG (H(IJ)), adds no informative data. The nested polytomy contains ambiguous information about relationships of areas CDEFG to each other and to areas HIJ. Areas A and B are basal to all other areas in the areagram of Figure 4.14, and so may be considered part

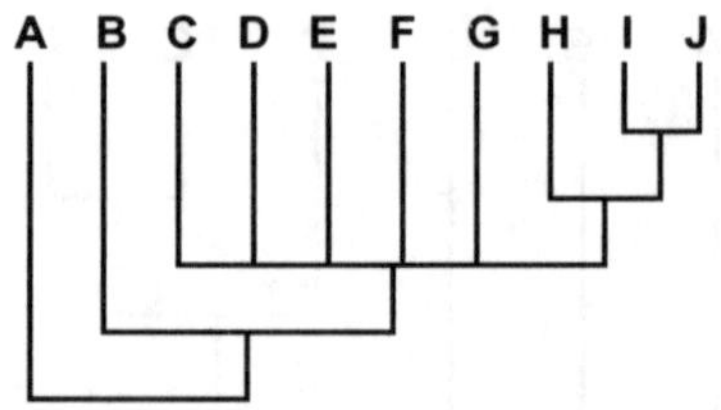

Figure 4.14. An example of a nested polytomy in an areagram. The single, informative subtree is H(IJ).

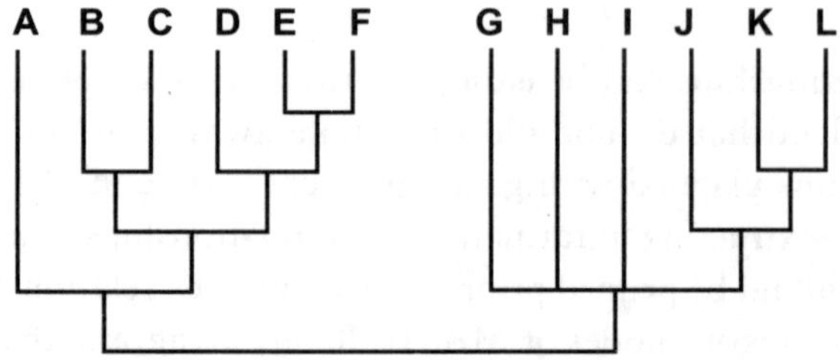

Figure 4.15. An example of a nested polytomy, G, H, I, and J(KL) in an areagram. There are only three informative subtrees: J(KL), A(D(EF)), and A(BC).

of the unresolved basal node. Because area relationships are added as one works "down" the areagram (from the terminals to the base—see Ebach et al., 2005), A and B are logically part of the unresolved series of relationships.

The procedure is the same when there is a polytomy nested within an areagram that has a series of resolved subtrees. For example, in Figure 4.15, although there is a nested trichotomy, there are three informative subtrees: J(KL), A(D(EF)), and A(BC). A is not included in a subtree with J(KL) because there is no close, resolved relationship between A and J(KL). As in the above examples, the nested polytomy contains ambiguous information about the relationships of areas G, H, and I to each other and to areas J, K, and L.

Crown Polytomies

Areagrams with polytomies at their crowns are ambiguous with regard to area relationships, even if there are basally resolved nodes. In the hypothetical areagram in Figure 4.16, for example, the polytomy EFGHIJ is uninformative with respect to area relationships because the area relationships are unresolved: there are no unique statements that can be made about any of these areas relative to the rest of the areagram.

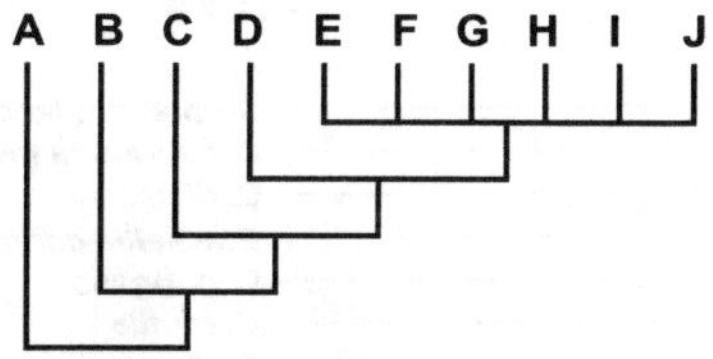

Figure 4.16. An example of a crown polytomy in an areagram. See text for area relationships it may contain.

One could argue that if the taxa in areas E, F, I, and J had gone extinct, we could identify a completely resolved set of area relationships: A(B(C(D(GH)))). True. But because these taxa are extant, the terminal polytomy EFGHIJ reflects ambiguous information about the relationships of their areas to each other and to the more basal areas.

The area relationships in the areagram of Figure 4.16 could be expressed in the following 14 sets of relationships:

A(B(C(D(EF)))) A(B(C(D(EG)))) A(B(C(D(EH)))) A(B(C(D(EI))))
A(B(C(D(EJ)))) A(B(C(D(FG)))) A(B(C(D(FH)))) A(B(C(D(FJ))))
A(B(C(D(GH)))) A(B(C(D(GI)))) A(B(C(D(GJ)))) A(B(C(D(HI))))
A(B(C(D(HJ)))) A(B(C(D(IJ))))

Although completely resolved, when added together these area relationships will always recover the unresolved polytomy of Figure 4.16.

Real areagrams show a range of polytomies. There are several examples of ambiguous polytomies (Figure 4.17a) in the areagram of four genera (*Trymalium, Pomaderris, Siegfriedia,* and *Crytandra*) of the tribe Pomaderreae of the global plant family Rhamnaceae (Ladiges et al., 2005). The areagram demonstrates relationships among areas of Australia, including Tasmania, and New Zealand, as mapped in Figure 4.17b.

Here, we focus on the areas from the south–west interzone and desert and from southwestern Australia, areas I, H, E, D, W, and G. From the areagram of Figure 4.17a, we extract the subtrees that do not have polytomies. There are three of these from our chosen region, each of which has at least one MAST (Figure 4.18a, left-hand column). There are seven resolved subtrees—that is, seven informative subtrees specified by treating the areas in the MASTs as separate occurrences (Figure 4.18a, right-hand column). These informative resolved subtrees may be added together to give a minimal tree specifying the relationships among areas I, H, E, D, W, G, and S (Figure 4.18b).

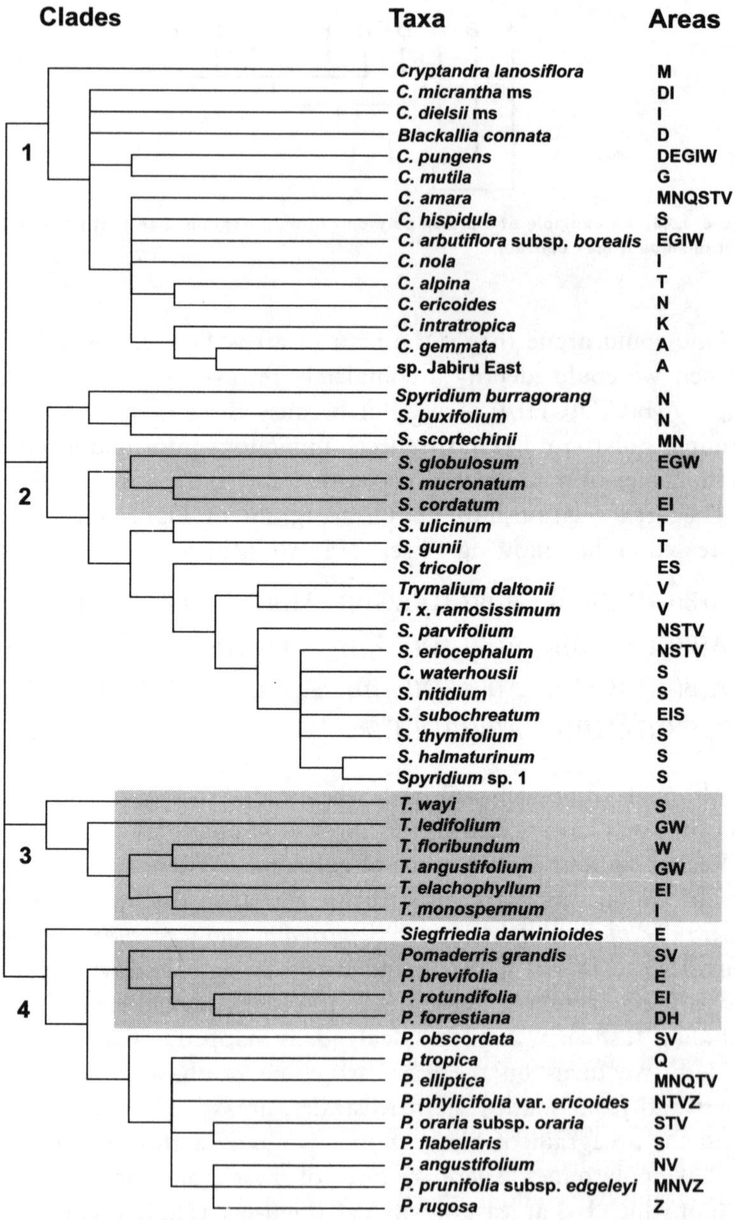

Figure 4.17a. Cladogram and areagram of the tribe Pomaderreae (after Ladiges et al., 2005: Fig. 2). A single letter abbreviation denotes each area (as in Figure 4.17b). The cladogram and areagram include many polytomies and relationships among the four major clades (1–4) that are unresolved. Informative subtrees that do not include polytomies are shaded (see also Figure 4.18a).

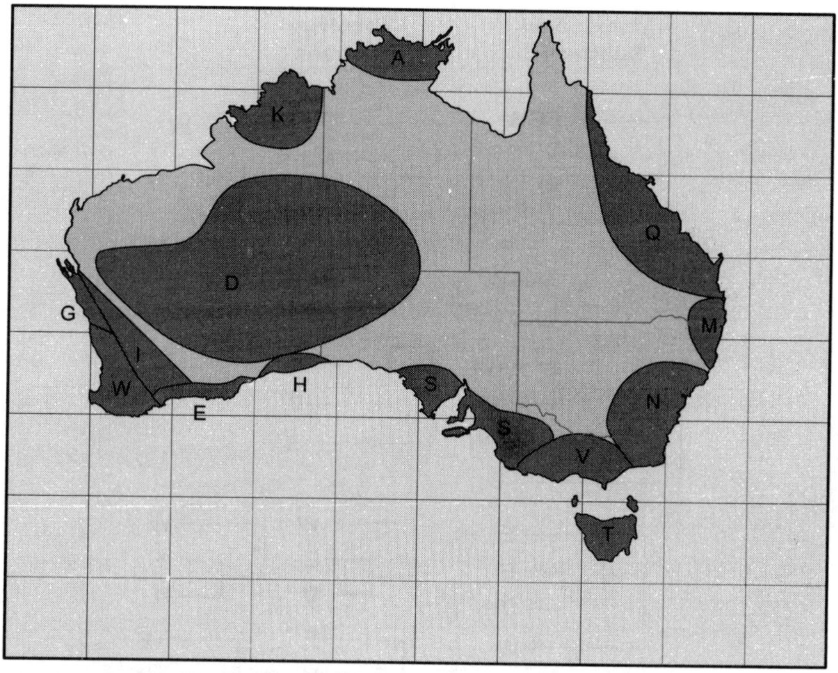

Figure 4.17b. Geographical areas of the tribe Pomaderreae in Australia (modified from Ladiges et al., 2005: Fig. 3). Note that "Z" refers to New Zealand, which is not shown on the map.

Using all relationships in the areagram, Ladiges et al. (2005) found five informative subtrees, 1 through 5 (Figure 4.19a), which, when combined, yield a minimal tree (Figure 4.19b). The result for the areas of interest in the Southwest and in the south–west interzone and desert—I, H, E, D, W, and G (Figure 4.19b, node 2)—is completely congruent with our result (Figure 4.18b).

It is not surprising that the result obtained by combining just the informative nodes is the same as that discovered by Ladiges et al. (2005) using all relationships. Informative relationships will always recover patterns, provided that they do not conflict. By including polytomies in areagrams, we could inadvertently add artifactual relationships. This does not happen in the Pomaderreae example because there are few paralogous MASTs—that is, MASTs that include areas found elsewhere in the areagram.

In Chapters 5 and 6, we review biogeographic methods, focusing on those that have been proposed to incorporate phylogenetic analysis

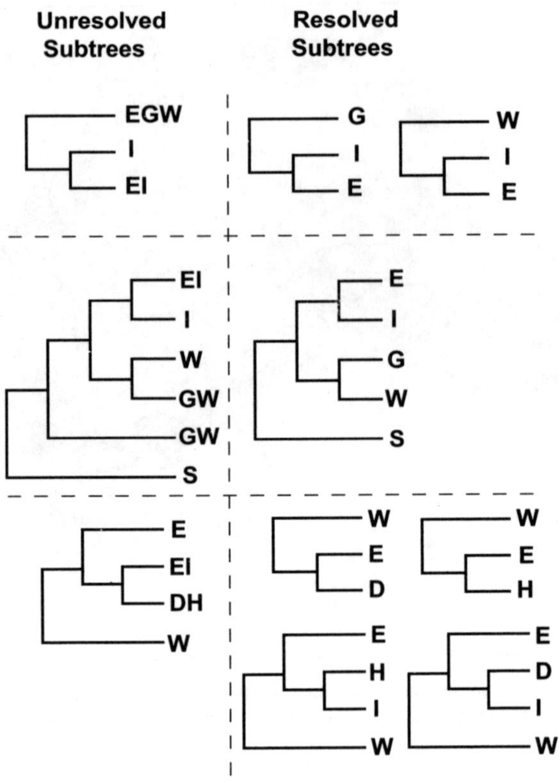

Figure 4.18a. Informative subtrees in the areagram of Figure 4.17a that do not include polytomies.

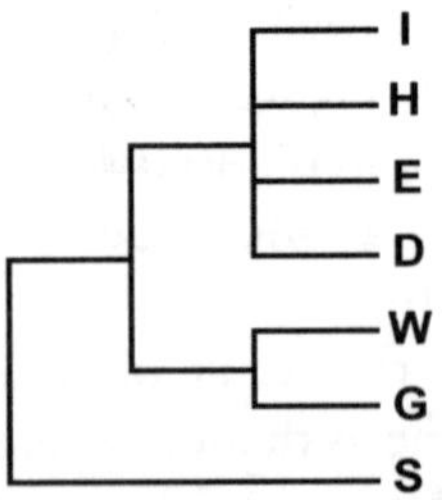

Figure 4.18b. The minimal tree for the tribe Pomaderreae, a combination of the seven resolved subtrees of Figure 4.18a (right-hand column).

into biogeography, and evaluate how well each method adheres to the principles presented here and in previous chapters. In Chapter 7, we outline a systematic biogeographic method for proposing biogeographical classifications based on geology, geography, and biological distributions and relationships.

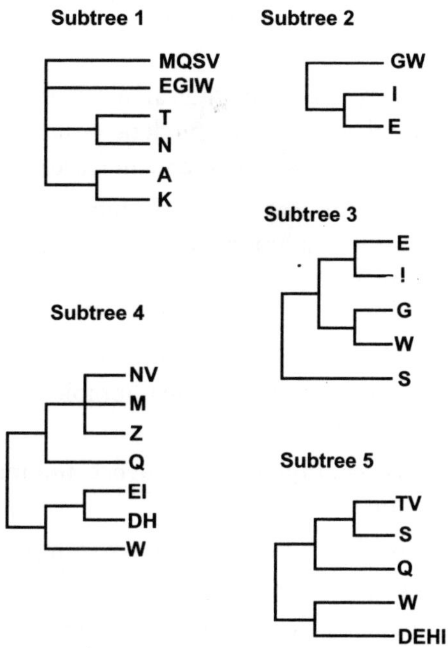

Figure 4.19a. The five informative subtrees (subtrees 1–5) for the tribe Pomaderreae as inferred by Ladiges et al. (2005: Figs. 4 and 5).

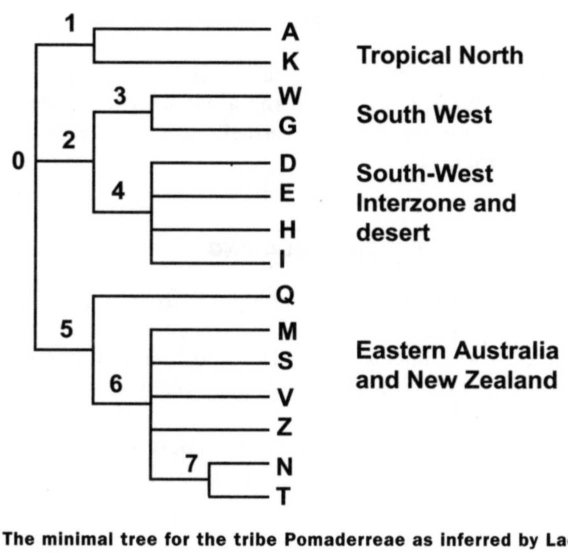

Figure 4.19b. The minimal tree for the tribe Pomaderreae as inferred by Ladiges et al. (2005: Figs. 4 and 5).

SUMMARY

- The role of biotic areas in biogeography is like that of taxa in systematics.
- A biotic area consists of homologous area relationships expressed by more than one monophyletic group that inhabits a common place and/or a designated endemic area.
- An area homolog, or three-area relationship, is the smallest unit of meaningful relationship among areas. Geographic paralogy, the repetition of areas on an areagram, provides no information on area relationships.
- Area homology, or area monophyly, is discovered when different areagrams are compared and their area homologs corroborate those of other areagrams.
- Area monophyly is geographical congruence and represents a natural classification.
- A single geographical area may contain two temporally overlapping biota.
- Unresolved components in areagrams are uninformative and cannot be resolved. Areagrams that contain deeply nested polytomies express no area homologs.

NOTE

1. A global area nomenclature will be based on a hierarchical classification of biotic areas—based on organisms and the areas in which they live. It will not be the "geography of names" (sensu Vermeij, 2002:935) that characterizes much descriptive biogeography.

FURTHER READING

Fitch, W. M. 1970. Distinguishing homologous from analogous proteins. *Systematic Zoology*, 19, 99–113.

Patterson, C. 1982. Morphological characters and homology. In K. A. Joysey and A. Friday (eds.), *Problems of phylogenetic reconstruction* (pp. 21–74). Academic Press, London.

Williams, D. M., and M. C. Ebach. 2008. *Foundations of systematics and biogeography*. Springer, New York.

PART II

METHODS

FIVE

BIOGEOGRAPHIC PROCESSES

BIOGEOGRAPHIC PROCESSES

Biologists and geographers come to biogeography from a broad range of fields, naturally bringing with them discipline-specific methods, assumptions, and goals, plus language, usually in the form of jargon. Systematists have applied cladistics, one method used to discover phylogenetic relationships among organisms, to analyses of area relationships using an array of methods, called by a variety of names: vicariance biogeography, cladistic biogeography, historical biogeography, phylogenetic biogeography, phylogeography, and comparative phylogeography, among others. Other systematists, in contrast, document and interpret distribution patterns without relying necessarily on cladistic hypotheses, particularly in panbiogeography, although these systematists rely on a

Overview

The differences between explanatory mechanisms and discoverable patterns have led to disagreements and misunderstandings in biogeography. Confusion between biogeographical processes and mechanisms is chief among these divergences.

Another divergence is between vicariance and dispersal mechanisms. This is considered by some to be the most important dispute in biogeography. We disagree. Much of the language of biogeography is loaded with implied mechanisms and processes.

Vicariance and dispersal are poorly understood biogeographical concepts. Those who favor one mechanism over the other treat it as an empirical process, not as a hypothetical, testable mechanism.

The distribution of all taxa and biota define their endemic areas, which means that no taxon or biota, by definition, will be found outside of its natural area. Many organisms disperse, such as through seasonal migration or to reach spawning grounds. This is the natural movement of an organism throughout its range, and it may occur over vast distances. Isolation and subsequent evolution of lineages occurs when this movement is disrupted.

We introduce the difference between areagrams and taxon/area cladograms, or TACs. Areagrams are to TACs as cladograms are to phylogenetic trees. Treating areagrams as TACs in biogeography obscures information on area relationships.

biological classification.[1] Ecologists have applied a range of methods, often called simply biogeography, or conservation management policies known under the rubric of applied biogeography. These methods are introduced here, and those that focus on discovery of biogeographic patterns are defined and classified in Chapter 6.

Regardless of the traditions of individual fields of investigation, a biogeographic method applied should be appropriate to the question asked. Comparative biogeographers first discover and classify biogeographic patterns and then interpret the process by which those patterns may have been formed. Not all of the methods that we define and classify share that goal. Ultimately, many biogeographers aim to identify biogeographic processes, the evolutionary mechanisms that caused the distribution pattern. Some methods incorporate an explanatory mechanism—an assumed, yet unobserved process—into the description of a pattern, and thereby fail to keep the tasks of pattern and process

Box 5.1 The Language of Biogeography

The words used to describe biogeographic patterns may reveal an inherent bias about mechanism. *Barrier* and *boundary* are two words used to describe the geographical limits of organic distributions. A *barrier* limits the passage or movement of an organism from one area to another. A *boundary* is a border or limit that marks the extent of a distribution. *Barrier* implies a mechanism; *boundary* does not.

Further complicating the language of biogeography, *barrier* has been used in vicariance and dispersal models to refer to different mechanisms. Consider sister species A and B that live on either side of the North Atlantic Ocean. A vicariance explanation for the distribution may be that there was once a widespread species that became disjunct when the Atlantic Ocean was formed. The barrier is the formation of the Atlantic Ocean: it drove speciation. A dispersal explanation may be that individual organisms swam across the Atlantic Ocean—say, from east to west—and became established in North America. The barrier is the Atlantic Ocean, which was breached for an unknown reason. Hence, the paradox: the barrier was not a barrier.

Explanatory dispersal is dependent on an explanatory vicariance event. If an organism disperses across an inferred barrier—here, the North Atlantic—over which it cannot return, then that dispersal logically is vicariance. The barrier, once breached, is now acting as a barrier. Explanatory vicariance simply means that formation of a geographical barrier *causes* genetic isolation. Dispersal, as an explanatory mechanism, is vague. Only genetic isolation makes sense in the light of evolution, highlighting the power of explanatory vicariance.

identification separate. Still other methods claim to address the same objectives, yet, as we will describe, cannot actually address the questions they were designed to answer.

Patterns: Processes versus Mechanisms

Process is defined here as the geographical, geological, ontogenetic, physiological, behavioral, and molecular changes, functions, or actions that bring about a change or result that we may observe and study under natural and experimental conditions. Processes over time form patterns. We use patterns to reconstruct or retrodict former processes. Pattern and process are complementary and inform one another.

Mechanism is defined here as an undetected event or narrative that explains a pattern or a process. Mechanisms are hypothetical. The explanatory nature of mechanisms makes them immune to contradictory evidence. All observations may support any given mechanism and may even support contradictory mechanisms.[2]

An example may help illustrate the difference between process and mechanism. The law of gravity may be described through empirical observation. Attempts to explain gravity have generated numerous mechanisms, such as an object of greater density of mass attracts an object of lesser density. We can see gravity work, but we may not know exactly how it works, and we do not need to know how it works to rely on it. If we see a helium-filled balloon rise, rather than fall, we may seek an explanation. We can revise the mechanism proposed for gravity, given empirical observations, or can retain it. This difference between process and mechanism may seem trivial, but it is crucial for understanding comparative biogeography: processes form patterns (e.g., what goes up must come down, except for helium and hydrogen gases), whereas mechanisms do not.

In biogeography, some commonly used terms refer to both a process and a mechanism, because the difference between the two concepts has not been well appreciated. We review the meanings of two such terms, vicariance and dispersal, the interpretation of which has fueled a seemingly intractable debate in biogeography.

VICARIANCE AND DISPERSAL

The terms vicariance and dispersal are, at the same time, both descriptive and explanatory or interpretive. A *vicariant* distribution in biogeography is of two (or more) mutually exclusive, closely related taxa, clades, or biota. The disjunct distribution does not necessarily have an explanation: two sister clades live on either slope of a mountain range, for example. The inference that the formation of the mountain caused the clades to differentiate and become disjunct is an explanation that is secondary to the observed distribution of the sister clades. One explanation for the distribution is that vicariance has occurred; an alternative explanation is that some member of the species dispersed across the mountain and subsequently differentiated (Figure 5.1). The vicariant distribution itself is not an explanation: how the population of species has become vicariant can be explained

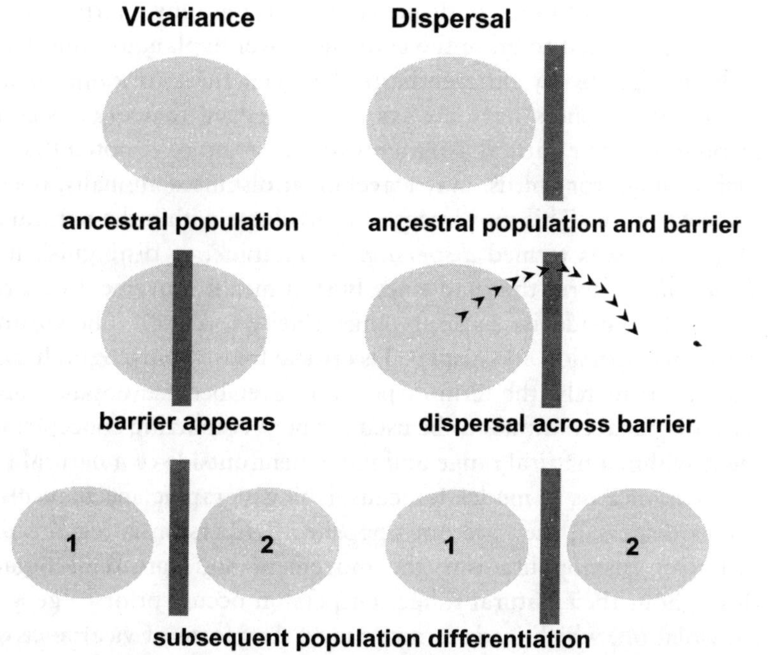

Figure 5.1. Diagrammatic representation of two possible causes for the same distribution of populations 1 and 2 on either side of a presumptive barrier (redrawn from Nelson and Platnick, 1984: Figure 1). Under the vicariance model, formation of the barrier caused a separation that led to differentiation of the populations; in the dispersal model, individuals migrated across the barrier, and this process was followed by subsequent population differentiation.

either by dispersal or by vicariance. The vicariant distribution can be explained in different ways.

In biogeography, dispersal has been used to describe almost any movement by an organism. Because organisms obviously move, dispersal has long been inferred to be the principal mechanism by which they became distributed. Dispersalism, or dispersalist biogeography, maintains that this mechanism is key: life *became distributed via various means of locomotion over a largely static Earth*. Because one of the traditional definitions of an endemic area is the area in which a taxon originated (see Chapter 3), the distributional mechanism has been edited to state, *"all clades were distributed by dispersal from a center of origin."* This movement is different from the movement of the organisms in the normal course of their lives.

There are two kinds of dispersal in biogeography. Norman Platnick (1976) attempted to erase the confusion over explanatory mechanisms in biogeography by differentiating between the two kinds of dispersal. Biogeographers have always acknowledged that organisms move throughout their natural geographical range and have noted that some, such as migratory birds, may travel great distances annually, over land and over water. This movement of individuals within the natural range of a taxon was termed *dispersion* by Platnick, to distinguish it from *dispersal*, the term that had since Buffon meant movement of a taxon, especially outside its natural range (Nelson, 1978). The distinction between dispersion and dispersal is critical for any biogeographic analysis. Unfortunately, the term dispersion never became popular, and the term dispersal continues to be used for both conflicting concepts: movement within a natural range and movement outside of a natural range.

Vicariance as a model for causal biogeographic mechanisms subsumes dispersal; they are not opposites. Dispersion is responsible for primitive cosmopolitanism, the movement and spread of organisms throughout their natural range. Dispersion occurs prior to geographical isolation, which leads to vicariance. Any form of vicariance occurs because of prior movement that established a broad range that was ultimately disrupted.

What of organisms that travel across large areas, such as across oceans or deserts, which are seemingly inhospitable, at least to us? An iguana that rafts from one Caribbean island to another has not dispersed if it has not left its natural distribution range. Shorefishes endemic to the Pacific Plate, as in the example in Chapter 3, are not dispersing if they swim from Enewetak to Bikini Atoll in the Marshall Islands. They are moving about their area of endemism, exhibiting dispersion, not dispersal.

The wrangle over dispersal versus vicariance models has caused a rift between those who want to use dispersal to explain *ad hoc*, usually one-way movement across inferred barriers and those who maintain that all geographical isolation results from vicariance. This is a weak rift because it is over explanations that each form the same distributional pattern. Further, vicariance as a mechanism operates at different levels. The splitting of a distribution range due to repeated earthquakes is vicariance; movement of a biota following a retreating glacier has also been called vicariance (see Platnick, 1976), but it may also be called dispersal—a holdover from pre-tectonic biogeography (see Adams, 1902). If vicariance and dispersal are ends of a continuum, then how did they come to be viewed as contradictory?

The strict differentiation between vicariance and dispersal was explicit in the explications of these models by Gareth Nelson and Norman Platnick, who implemented a fusion of Croizat's vicariance with Hennig's cladistics. The diagram of Figure 5.1, a version of which was first published in an educational manual, has been duplicated many times in many forms, in classrooms, papers (e.g., Crisci, 2001: Figure 1), and texts (e.g., Crisci et al., 2003: Figure 1.1). The diagram was meant to illustrate simply the relationship between biological distribution and Earth history, and furthermore, to demonstrate that the same distribution could have different explanations. They had earlier fully examined the relationship between vicariance and dispersal as potential explanations for distributions (e.g., Platnick and Nelson, 1978) and had argued that patterns, not explanations, should be sought, especially if it was not possible to choose among an array of possible explanations. They were careful to explain that if one were to try to choose between vicariance and dispersal as explanations by demonstrating that the age of the barrier was younger or older than the lineages, that one had not demonstrated evidence for either dispersal or vicariance, respectively. Instead, he had demonstrated that one estimate of the age of the lineages was older or younger than one estimate of the age of the presumptive barrier. We say "presumptive barrier" because even if two populations live on either side of a mountain range, it is a hypothesis, not an observation, that formation of that barrier had anything to do with differentiation of the populations. And as we have said from the outset, it is patterns of distribution as demonstrated by area homology, the relationships among three areas, that we seek. Populations 1 and 2 of Figure 5.1 specify no area homology; the diagram is only about mechanisms.

Despite these warnings, a popular exercise in biogeography, spurred on by increasingly sophisticated techniques to estimate ages of lineages via a molecular evolutionary clock, is to arbitrate between dispersal and vicariance as the cause of a distribution, often of just two areas. Ages of lineages and presumptive barriers are compared, as above, and the results are presented as *evidence* for either dispersal or vicariance (see Box 4.2). These comparative ages are not evidence, as they represent no more than estimates of the time of one event compared with estimates of the time of another. Insofar as they ignore area homology, they add little to our understanding of the historical distribution of taxa or of the relationships among areas.

Box 5.2 Nancy Tyson Burbidge (1912–1977)

Figure 5.2. Nancy Tyson Burbidge (Prints 095/16) http://www.anbg.gov.aulbiography/burbidge.biography.html [Copyright Centre for Plant Biodiversity Research, Australian National Botanic Gardens.]

Nancy Tyson Burbidge, Australian plant systematist, was born in Yorkshire, England, and immigrated as a child to Australia. She received her bachelor's, master's and doctoral degrees from the University of Western Australia.

Burbidge was a botanist at CSIRO, Canberra, from 1946 to 1973, reaching the position of Curator of the Herbarium. She is known for her comprehensive *The Phytogeography of the Australian Region* (Burbidge, 1960), among many other scientific papers and books. Her method was distinctly panbiogeographic (Ladiges, 1998), as she attributed development of the Australian flora to "migration by communities" rather than to chance, long-distance dispersal. Monuments to her include the Nancy T. Burbidge Memorial Amphitheatre at the Australian National Botanic Gardens in Canberra.

The confusion over *descriptive vicariance* and *explanatory vicariance* that we describe here has stemmed from those who mean the *process* versus those who mean the *mechanism* of vicariance.

Descriptive Vicariance and Dispersal

Descriptive vicariance is a way to *describe* allopatric or parapatic distributions. As descriptive vicariance is totally dependent on the taxic and/or biotic extent of areas, it does not *explain* which existing or extinct mechanisms are responsible for biotic and taxic isolation. It simply shows where biota or taxa are (Figure 5.1).

Descriptive dispersal (dispersion of Platnick, 1976) identifies the extent to which organisms are able to move within their natural range. If the area in which organisms move is interrupted by a barrier, then the two resultant parts of the biota have vicariated. *Because the distribution of all taxa and biota define their endemic areas, no taxon or biota, by definition, will be found outside of its natural area.* A biota may extend beyond the boundaries of an artificial "barrier," such as a state or provincial border, but it never leaves its own area. Each taxon is endemic to some area, large or small. The concept of area monophyly means that the limits of natural distributions need to be discovered, rather than artificial boundaries proposed. All artificial boundaries, because they designate unnatural areas, will circumscribe paraphyletic areas. If an organism disperses (moves or migrates) outside an inferred boundary, then our understanding of its natural distribution is extended. *Organisms, therefore, cannot disperse outside their own areas of distribution.* If two populations of one species of rodent, for example, live on two Caribbean islands, they might be assumed to be part of two separate areas. If several individuals are able to raft between the islands, then the area includes the two islands, despite the inferred barriers of water and island limits. Isolated groups—biota or taxa—may not necessarily be differentiated.

Explanatory Vicariance and Dispersal

Explanatory dispersal and vicariance are *ad hoc* mechanisms that explain distributions. They are unobservable and hypothetical. In the hypothetical example of the distribution of two sister taxa, A and B, on either side of the North Atlantic (Box 5.1), the movement of organisms outside their perceived area is interpreted differently. Under such models, organisms are assumed to be able to move away from their own endemic area. This is so because endemic areas are interpreted as *centers of origin* rather than as areas where the organism or biota live. Endemic areas, if defined by the biogeographer as artificial or hypothesized distributions, are rigid. Movement away from an endemic area is equal to movement across a predefined barrier or boundary (Figure 5.1). In evolutionary biogeography, the area may be redefined to include a new area into which the organism has dispersed; dispersal outside the predefined area may be termed a "dispersal event." In comparative biogeography, the definition of the area of endemism would be modified.

The most popular explanatory dispersal model is long-distance dispersal (see Mayr, 1982). Long-distance dispersal was invoked regularly in biogeography to explain distributions, either continuous or disjunct (i.e., vicariant), on an assumed static Earth. The belief in fixed continents and a slowly changing Earth meant that organisms alone were responsible for dispersal, which encouraged assessment of physiological dispersal ability and comparison of that with the relative penetrability of geographical barriers. The discovery of fossil disjunct plant- and land-dwelling mammals and reptiles led to rejection of long-distance dispersal by many 19th-century biologists, such as Joseph Dalton Hooker, who lived under the prevailing geological hypothesis of a static Earth. The continents, not the organisms, must have been dispersing or the land sinking. The acceptance of continental drift (see Chapter 1) and a mobilist Earth gave life to vicariance models; they were no longer incredible, but plausible (e.g., Leviton and Aldrich, 1986).

Long-distance dispersal as an important biogeographic mechanism has become well established in distributional thinking. Many defend long-distance dispersal with arguments that are the same as those of the pre–continental drift era: the organisms are too young to have been affected by major patterns of Earth history. Long-distance dispersal requires organisms to have the physiological ability to disperse over great distances in a short period of time: jump-dispersal, according to Pielou (1979), a mechanism that allows the establishment of a new population from a few successful migrants.[3] This type of dispersal can explain any and all distributions. It is also not comparative and therefore does not incorporate concepts such as "migration by communities" (see Box 5.2), explanations that are at the same time vicariant and dispersalist.

OTHER EXPLANATORY MODELS

The systematic biogeographic method—focused on discovering area homology and congruence—ultimately has great explanatory power as it subsumes all possible explanatory models, both vicariant and dispersalist. Here, we describe three models that have been invoked to explain area patterns or portions of patterns.

Biotic Dispersal

Vicariance and dispersal come together in the model of *biotic dispersal,*[4] a term coined by Platnick and Nelson (1978) to refer to the dispersal of several elements of a biota into a previously unoccupied area, promoting cosmopolitanism, or widespread taxa that may subsequently be disrupted by vicariance. The concept was well understood by Wallace and his contemporaries. Some major events in Earth history that facilitated biotic dispersal include the closing of the Tethys Sea, with juxtaposition of the Asian and Australian parts of the Indo-Australian archipelago, and the movement of the Caribbean tectonic plate to its current position east of the Panamanian isthmus. Such large-scale dispersal of biota (taxa and areas) can also be described as large-scale vicariance. Biotic dispersal has been one of the most important mechanisms of biogeographic distributions.

Extinction

Extinction has played a critical, yet poorly understood, role in the formation of biotic distributions. Extinction may help explain why some taxa are missing from areas where biogeographers expect to find them. Why are there no endemic African cacti? Why is the Tasmanian tiger, the thylacine *(Thylacinus cynocephalus),* absent from mainland Australia? Fossils demonstrate that thylacines once lived on the mainland but have gone extinct. Widespread extinction of the Earth's biota since the Upper Mesozoic has left fragments or pieces of biogeographic patterns. Absence of a single taxon cannot form a pattern, although fossil evidence can add areas to an analysis and allow discovery of area homology.

Extinction, especially mass extinction, can be correlated with other inorganic processes, such as sea-level rise, continental drift, and bolide impacts, just to name three. Extinction mechanisms, like all other mechanisms, are not of primary concern to biogeographers because the same pattern could be formed regardless of whether a taxon went extinct because it was hit by an asteroid or because it was eaten by a competitor. Extinction (loss of taxa over time) is a mechanism in biogeography, but as extinction does not form patterns, it is not a biogeographic process.

Ecological Stranding

Ecological stranding is a mechanism that results in changes in the habitat of a biota over time without requiring movement or dispersal of members of the biota. Events of Earth history can cause one portion of a biota to go extinct while another portion lives on. If the biota lives in a varied habitat, one that is transitional between marine and freshwater, and if the marine habitat is obliterated, the biota may persist in freshwater. Isolation in one habitat or the other has been termed *ecological stranding* (Craw et al., 1999). Isolation, or stranding, through changes in sea level or altitude can result in a former euryhaline species being restricted to freshwater, for example. Such isolation does not require or imply invasion of that habitat, although such isolated taxa are often described as having "invaded" freshwater. There is no evidence for extraordinary movement or dispersal: some taxa persist in one habitat, others in another habitat.

Ecological stranding is one of the most important biogeographic mechanisms throughout geological time, to the present day. Regression of epicontinental seas in North America and Europe, for example, obliterated the shallow marine habitat. Portions of the coastal or intertidal biota that survived became inland biota simply by staying in place.

WHY NOT TO OPTIMIZE AREAS IN BIOGEOGRAPHY: AREAGRAMS VERSUS TAXON/AREA CLADOGRAMS (TACS)

Just as the language of biogeography may be loaded with implied process or mechanism (Box 5.1), so may the methods. One method in particular that has been borrowed from phylogenetic systematics that has no place in biogeography is the optimization of areas on the nodes of an areagram. The method was formulated by Hennig (1966) as the *Progression Rule* and was adopted by Brundin (1966, 1972) and many other phylogenetic biogeographers. The Progression Rule was based on Hennig's assumption that the phylogenetically primitive members of a lineage would be found near the center of origin whereas progressively more derived members would have moved away from that center. Optimization of areas was interpreted as a "road map" for the distribution of the lineage.

Despite its simple appeal, optimization has several theoretical and methodological limitations that make it inappropriate as a method in

Box 5.3 Why Not to Optimize Areas on Areagrams

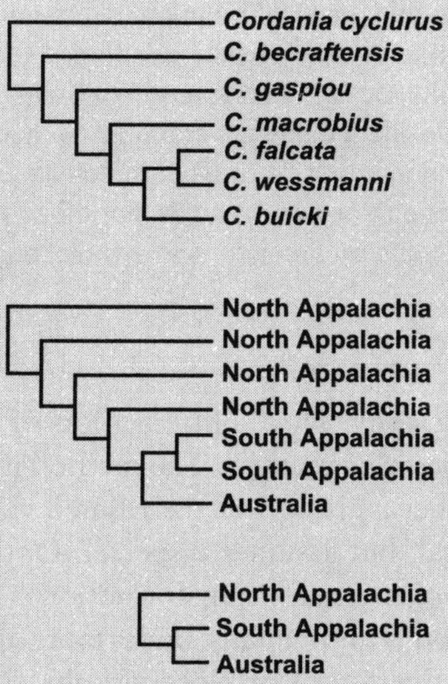

Figure 5.3.Cladogram (above) and areagram (middle) of the Devonian trilobite *Cordania* (following Ebach and Edgecombe, 1999: Figure 2). The areagram specifies one area homolog: North Appalachia (South Appalachia, Australia).

Single biogeographical distributions do not support any particular evolutionary processes, patterns, or mechanisms. Given this, we examine the cladogram of the Devonian trilobite genus *Cordania,* as proposed by Ebach and Edgecombe (1999: Figure 2).

The *Cordania* species live in three areas: North Appalachia, South Appalachia, and Australia. We can convert the *Cordania* cladogram into an areagram or taxon area cladogram (TAC, see Chapter 7) by replacing the names of the taxa with areas in which they live.

All event-based explanations, such as sympatry, vicariance, dispersal, and so forth, are inferred by relationships among areas and by geographic paralogy or MASTs. By adding *geographical data* onto cladograms, the resulting TAC or areagram is assumed by some methods to contain information about evolutionary mechanisms *prior* to using the TAC or areagram to search for a pattern or geographical congruence (see Chapter 6).

There is one area homolog specified by the areagram: North Appalachia (South Appalachia, Australia). There is no geographic congruence, as we have just one area homolog; other taxa must be examined to assess if this is a general statement of area relationships or one specified solely by these species of *Cordania*. The area North Appalachia is *(continued)* geographically paralogous, or repeated, on the areagram, but this repetition provides no additional information on area relationships. Optimizing areas on the areagram—that is, inferring that *Cordania* originated in North Appalachia and dispersed to South Appalachia and Australia—means invoking a dispersalist model (without a test of the model), and furthermore, treats the areagram not like a cladogram but like a phylogenetic tree (see Chapters 4 and 7). For these reasons, we never advocate optimizing areas on areagrams in any biogeographic analysis.

comparative biogeography. First, optimization of areas is an explicit center of origin, dispersalist method. The method implies a mechanism. It does not ask whether a group was distributed via the mechanisms of vicariance or dispersal, but assumes dispersal. Optimization assumes a transformation; in biogeography, transformation is movement from one area to another. Second, and equally important, optimization of areas ignores the comparative aspect of biogeography. Rather than optimizing areas, biotic area homologies should be extracted and then tested to see if they are general or unique. Finally, optimization of areas confuses areagrams and taxon/area cladograms, or TACs. A taxonomic cladogram in which the names of taxa have been replaced by the areas in which they live may be interpreted as an areagram or TAC, comparable to taxon cladograms and phylogenetic trees, respectively, in systematics. Areagrams contain solely information on area homologies; TACs are interpreted to convey information on distributional mechanisms, centers of origin, and so on.

We explore the difference between areagrams and TACs more fully in Chapter 7.

Other methods of biogeography will be reviewed in Chapter 6. Our goal is not an exhaustive critique of biogeographic methods, but rather an appraisal of how well they meet the aims of a comparative biogeography.

SUMMARY

A process is a geographical, geological, ontogenetic, physiological, behavioral, or molecular change, function, or action that brings about a result we can observe and study under natural and experimental conditions.

- A mechanism is an undetected event or narrative that explains a pattern or a process.
- Vicariance has greater explanatory power and includes dispersal.
- Areagrams contain information on area homologies.
- Taxon/area cladograms (TACs), like phylogenies, are interpreted with respect to distributional mechanisms.
- Ecological stranding is one of the most important biogeographic mechanisms: it results in changes in the habitat of a biota over time without requiring movement or dispersal of members of the biota.

NOTES

1. Crisci (2001) recognized nine basic historical biogeographic approaches: center of origin and dispersal; panbiogeography; phylogenetic biogeography; cladistic biogeography; phylogeography; parsimony analysis of endemicity (PAE); event-based methods; ancestral areas; and experimental biogeography. These do not all necessarily require a phylogenetic approach, nor are they all comparative.

2. The debate that raged in systematics over pattern versus process (see Hull, 1988) could be reconsidered as a debate over pattern versus mechanism.

3. Accidental transport of invasive species, through ballast water, for example, is human-mediated dispersal. Distribution of such taxa requires no comparative biogeographic analysis (see also Chapter 3).

4. Biotic dispersal has also been called *geodispersal* (Lieberman and Eldredge, 1996). Geodispersal is a more recent term that is synonymous with *biotic dispersal,* which has precedence. See discussion in Parenti (2006).

FURTHER READING

Lomolino, M. V., and L. R. Heaney, (eds.). 2004. *Frontiers of biogeography: New directions in the geography of nature.* Sinauer Associates Inc., Sunderland, Massachusetts.

Roughgarden, J. 1995. *Anolis lizards of the Caribbean: Ecology, evolution and plate tectonics.* Oxford University Press, New York.

Spellerberg, I. F., and J. W. D. Sawyer. 1999. *An introduction to applied biogeography.* Cambridge University Press, Cambridge.

SIX

BIOGEOGRAPHIC METHODS AND APPLICATIONS

COMPARING BIOGEOGRAPHIC METHODS AND APPLICATIONS

An explosion of biogeographic methods and applications in the late 1970s and early 1980s coincided with a numerical revolution in systematics (see Crovello, 1981; Crisci, 2001; Williams and Ebach, 2008; Fattorini, 2008). Biogeography adopted many methods from cladistics, phylogenetic systematics, phenetics, and ecology—many of these matrix based and incorporating parsimony algorithms. The molecular revolution that followed in the 1990s also influenced biogeography, although few particular methods

TABLE 6.1 A CLASSIFICATION OF BIOGEOGRAPHIC METHODS

1. Systematic Biogeography
- Descriptive Biogeography
  - Distribution Maps (sensu Zimmermann, 1777)
  - Biogeographic Maps (sensu Lamarck and Candolle, 1805)
  - Panbiogeography (Croizat, 1964; Craw et al., 1999)
- Cladistic Methods
  - Component Analysis (Nelson and Platnick, 1981; Nelson, 1984)
  - Assumptions 1 and 2 (Nelson and Platnick, 1981)
  - Paralogy-free Subtree Analysis (Nelson and Ladiges, 1996)

2. Evolutionary Biogeography
- Ecographic Methods
  - Niche Modeling (Peterson et al., 1999)
- Phylogenetic Methods
  - Phylogenetic Biogeography (Hennig, 1966; Brundin, 1966)
  - Phylogeography (sensu Avise, 2000)
  - Phylogenetic Analysis for Comparing Trees (PACT) (Wojcicki and Brooks, 2005)
- Matrix-Based Methods
  - Parsimony Analysis of Endemicity (PAE) (Rosen, 1985, 1988)
    - *Primary and Secondary PAE (Rosen, 1988)*
    - *Cladistic Analysis of Distributions and Endemism (CADE) (Porzecanski and Cracraft, 2005)*
  - Brooks Parsimony Analysis (BPA) (Wiley, 1987)
    - *Assumption 0 (Zandee and Roos, 1987)*
    - *Comparative Phylogeography (Riddle, 2005; Riddle and Hafner, 2004)*
  - Ancestral Area Analysis (AAA) (Bremer, 1992, 1995)
  - Dispersal and Vicariance Analysis (DIVA) (Ronquist, 1997)

were developed specifically for a molecular-based biogeography. We group biogeographic procedures into two general types: *methods* (Table 6.1) and *applications* (Table 6.2). Our goal is not to present an exhaustive review of methods and applications in biogeography—something that is well beyond the scope of this book. Instead, we focus on methods and applications that have as their primary goal the discovery and explanation of biogeographic patterns through time.

Overview

Biogeographical methods and applications are abundant, vary in their aims, and are classified in different ways. We contrast methods and applications in biogeography by aim: to describe and classify biotic areas and discover general patterns or to explain a biotic distribution through a series of hypothetical mechanisms.

Our classification attempts to place the most popular of these into systematic (descriptive and cladistic) and evolutionary (ecographic, phylogenetic, and matrix-based) biogeographical methods and applications. We provide a short description of each.

Cladistic biogeography has been used to describe a range of methods, some of which analyze matrices of presence/absence data of taxa in areas and may also interpret areagrams as taxon/area cladograms (TACs). We restrict the term *cladistic biogeography* to those methods that convert taxon cladograms into areagrams, not into TACs, and that infer biotic area homologs from such areagrams.

SYSTEMATIC BIOGEOGRAPHIC METHODS

Systematic biogeographic methods are either purely descriptive—observations or records of data—or they aim to discover biogeographic patterns. Discovering patterns does not require postulating explanations, evolutionary, ecological, or geographical, prior to analysis. Proposing explanations for patterns is the goal of evolutionary biogeographic methods. Systematic biogeographic methods provide the data on the spatial arrangement of life that allow the evolutionary, ecological, and geographical explanations to be made.

Descriptive Methods

Descriptive biogeographic methods rely on taxonomy, include the known distributions of taxa, and may also use ecological, geographical, or geological data to infer a range: the space occupied by a taxon through time. Descriptive methods do not rely on phylogenetic data or particular evolutionary models. The goal of descriptive methods is straightforward: to list or map taxic distributions or to classify biotic areas based on observed or inferred geographical or ecological boundaries alone or in combination with the distributional limits of taxa, fossil and living. No existing empirical tests assess if descriptive

TABLE 6.2 A CLASSIFICATION OF BIOGEOGRAPHIC APPLICATIONS

Systematic Biogeographic Applications
- Taxonomic Techniques
  - Hierarchy Analysis (described in Chapter 7)
- Geographical Techniques
  - Terrane Analysis (Young, 1987)
  - Area Cladistics (Ebach, 1999)
  - Biogeographic Reconstruction (sensu Glasby, 2005)
  - Hovenkamp Vicariance Analysis (HVA) (Hovenkamp 1997, 2001; Fattorini, 2008)

Evolutionary Biogeographic Applications
- Geological Techniques
  - Paleogeography
- Temporal Techniques
  - Stratophenetics (Gingerich, 1979) and Stratocladistics (Fisher, 1994, 2008)
  - Temporal Geographic Paralogy (Zaragüeta et al., 2004)

biogeographic classifications are natural; they require no evolutionary models to explain taxic distribution. Descriptive methods, by definition, are associated with taxonomic, ecological, or geographical accounts; formal taxonomic description of a new species usually includes a statement on its distributional range, or, at least the type locality. These methods provide some of the most basic, raw data of biogeographic analyses. When mapped, the data provide a visual representation of a distribution pattern or patterns that can be compared to others.

Distribution Maps

Distribution maps are among the most fundamental and historically informative data of any biogeographic study (Figure 6.1a). The majority of maps used in biogeography are distribution maps. The first such map, as far as is known, was drawn by Zimmermann (1777) to show the distributions of Asiatic and African elephants and camels, as well as the localities of certain other vertebrates (see Chapter 2). Any cartographic representation of taxic distributions (e.g., a range map), taxon localities, taxic movements (i.e., dispersal, migration, or dispersion), or hypothetical barriers is a distribution map (e.g., Schmarda, 1853; see below). A taxon distribution map may reveal historical distribution

Box 6.1 Geographical Information Systems: GIS

Geographical information systems, or GIS, store geographically referenced information in databases that can be searched and analyzed for trends in geographic data. Sophisticated mapping programs (e.g., ArcView®) that incorporate GIS technology are being used increasingly to predict and analyze taxic distributions based on known habitat and diversity values.

The field of ecography, in particular, uses GIS in association with mapping programs or statistical packages to depict fine-scale point distribution data and to estimate distribution ranges and species abundance.

GIS data provide distributional data with which to draw comparisons between species based on ecological or geographical attributes. Various programs are available that model the ecological niche of a species to predict, for example, where in the world an invasive species would most likely become established. When combined with phylogenetic information, such technologies allow biogeographic studies to operate at extremely fine and detailed scales.

clues when compared with other cartographic representations (i.e., geological, geographical, or oceanographic maps); potential geographical limits (a mountain, seaway, or margin of a tectonic plate); or potential dispersal routes (direction of an ocean current), for example.

Biogeographic Maps

Biogeographic maps are classificatory schemes used in biogeography to define, and often to compare, biotic regions or endemic areas. Biotic regions and the information they convey on geography, geology, climatology, ecology, soil chemistry, and biotic limits (e.g., Wallace's Line) through time can be represented on biogeographic maps. These maps differ from distribution maps in that they do not necessarily detail specific information regarding a single taxon, such as the summer and winter migratory ranges of swallows, but instead focus on biota. The first biogeographic map, using this definition, was drawn by Lamarck and Candolle (1805), and the term *biogeographic* was coined by Ratzel (1891; see Chapter 2). Biogeographic maps involve biogeographic analysis (Figure 6.1b) and may be augmented by explanatory text describing and classifying the areas depicted.

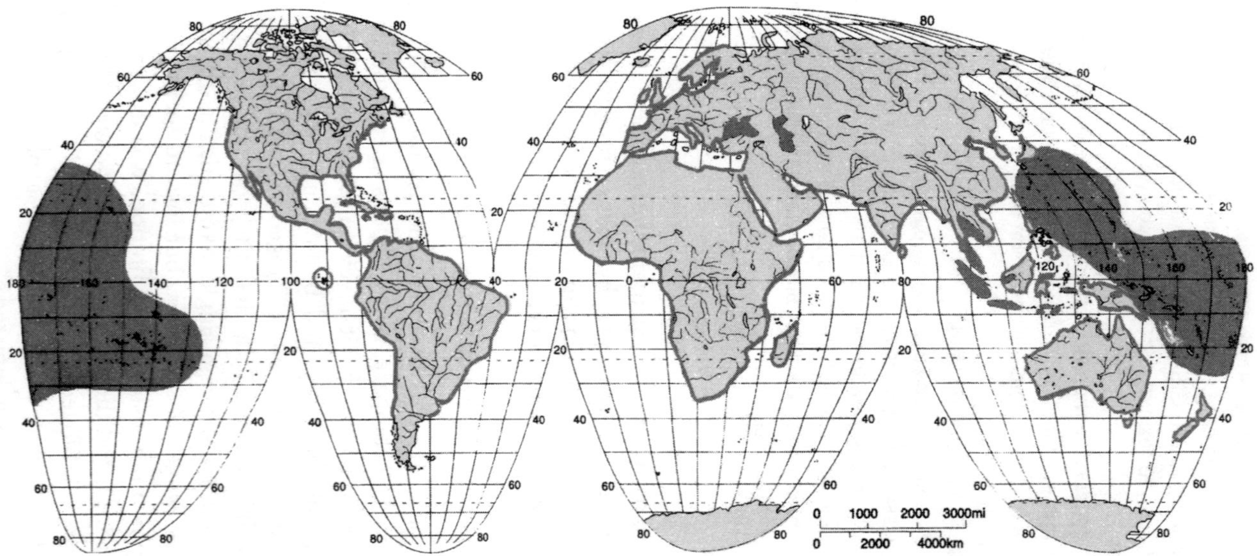

Figure 6.1a. Distribution map. The approximate distribution and distributional limits of pipefishes, family Syngnathidae (modified from Berra, 2001:355). [Map copyright Tim Berra.]

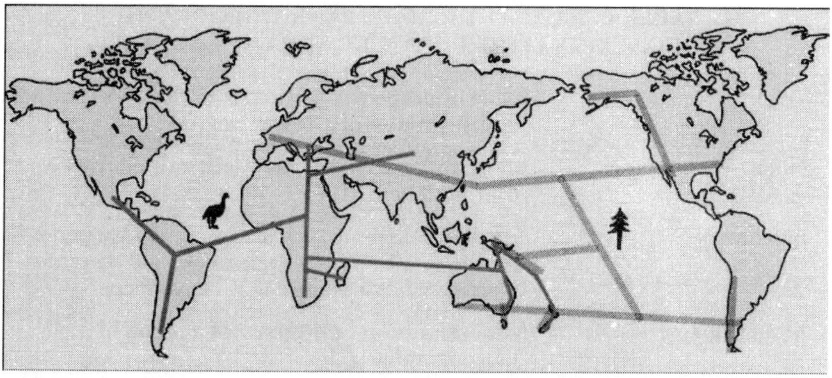

Figure 6.1b. Biogeographic map. Panbiogeographic tracks (from Page, 1989: Figure 10) demonstrating that ratite birds (solid lines) and southern beeches (hatched lines) have different baselines and are not panbiogeographic homologs (following Craw, 1985). [Modified map reproduced with permission of Carolyn King.]

Panbiogeography

Biogeographic maps figure prominently in the theory and methods of panbiogeography. Léon Croizat developed *track analysis* as the cornerstone of panbiogeography. Tracks are drawn on maps to show the biological connections between disjunct areas; their depiction is, figuratively, a biogeographical "connect-the-dots" method. Croizat was not the first biogeographer to draw tracks. This method had already made its way into texts by the early part of the 20th century (Box 6.2), but Croizat became linked with the track concept, as he was the principal proponent of its development as vicariance theory evolved (see Chapter 1).

A track is one of the four main concepts of panbiogeography (Table 6.3). Tracks allow us to see how disjunct distributions are spatially connected (Figure 6.1b). A track is not strictly a distribution, but a connection of localities across space. Croizat never specified a method for recognizing or drawing tracks on maps; tracks could be rendered as solid or broken lines to connect localities or as circles to denote the limits of a distribution, for example. It fell to a group of New Zealand panbiogeographers, including Robin Craw, John Grehan, Michael Heads, and Roderic Page, to formalize many of Croizat's methods. Minimum spanning graphs were used to connect distribution records to form tracks.[1] A *main massing* is a concentration of diversity of a taxon in a biogeographic space and may be interpreted as a "center of diversity."

TABLE 6.3 THE FOUR MAIN PANBIOGEOGRAPHIC CONCEPTS (AFTER CRAW ET AL., 1999:20–22)

Track	A line or graph drawn on a map that links the areas of distribution of a taxon or group of taxa
Node	An area where two or more generalized tracks overlap
Baseline	A geological feature, such as an ocean basin, common to more than one track, and therefore interpreted as a feature that unites them
Main Massing	A concentration of diversity of a taxon in a biogeographic space

When more than one track of unrelated or distantly related taxa overlap, they form a *generalized track,* also called a *standard track.* Generalized tracks have been interpreted as representing ancestral distribution areas. They are significant also for demonstrating the highly repetitive elements inherent in biological distribution worldwide.

Two or more generalized tracks intersect at a panbiogeographic *node.* Nodes may be characterized by high degrees of endemism. Croizat recognized five major nodes, numbered 1 through 5, among a global network of generalized tracks (Figure 1.1); he did not indicate how they might be related hierarchically, but he did relate them to geological features. The track network is analogous to area relationships drawn as a reticulation or polytomy: all possible relationships are supported.

Tracks are oriented spatially to span baselines. A *baseline* is a geological feature, such as an ocean basin or mountain chain, that is marked by several tracks and interpreted as a unifying biogeographic concept: Pacific Basin endemics, Atlantic Ocean endemics, and so on. Tracks that share a baseline may be *spatial homologs,* in a phenetic, not cladistic, sense. Tracks that do not share a baseline are not homologs, in either panbiogeography or cladistic biogeography. The flightless ratite birds and southern beeches of the genus *Nothofagus* are distributed throughout Gondwana and have been considered, therefore, to share a history. Demonstration that tracks of the ratites and southern beeches span different baselines (Figure 6.1b) led Craw (1985) to conclude that they are not panbiogeographic homologs.

Panbiogeography, especially track analysis and mapping, benefits systematic biogeography in illustrating repeated, overlapping spatial distributions. In a monograph on a portion of the Ericaceae, the heath family, Heads (2003) used track analysis to illustrate disjunct floral

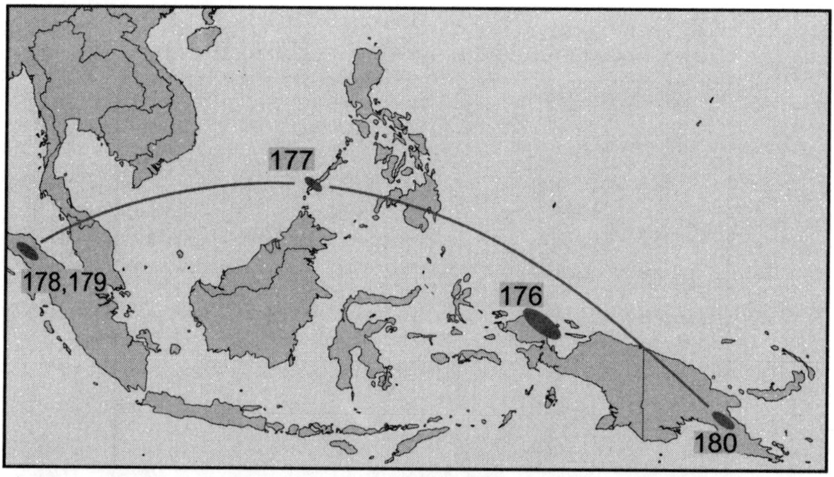

Figure 6.2. Distributional tracks in *Rhododendron* ser. *Buxifolia* III (modified from Heads, 2003: Figure 24). The Vogelkop Peninsula (New Guinea) species (176) is sister to a group of four species (177 to 180) that demonstrate a "vicariant arc" (Heads, 2003:362).

distributions across Malesia, a broad floristic province that includes the Malay Peninsula, the Philippines, Indonesia, and New Guinea. The *Rhododendron* ser. *Buxifolia* III species group is broadly distributed across Malesia in Sumatra, the Philippines, and New Guinea (Figure 6.2). The disjunct or vicariant distributions of four species are joined together to illustrate their taxonomic relationship. They are sister to the New Guinea species, a relationship not shown in the mapped tracks.

Tracks may also join biota or areas that share similar terranes, geological or geographical. The Ericaceae genera *Vaccinium* and *Rhododendron* are "closely associated with active or recently active volcanoes and their highly disturbed environs" (Heads, 2003:337). Should we survey *similar* environs or habitats within Malesia, we would predict that these or other ericaceous plants live in these habitats. Panbiogeography is a method of *similarity*, which uses common elements (taxa or common terranes) to relate areas that share the same biota or the same ecology. This relationship is not hierarchical; the relationship of one track to a second relative to a third is not considered.

Cladistic Biogeographic Methods

Many biogeographic methods use cladograms of taxa as the starting point for interpreting distributions, but we do not classify all of them as cladistic, nor do we consider all of them to be phylogenetic. A

Box 6.2 Marion Isabel Newbigin (1869–1934)

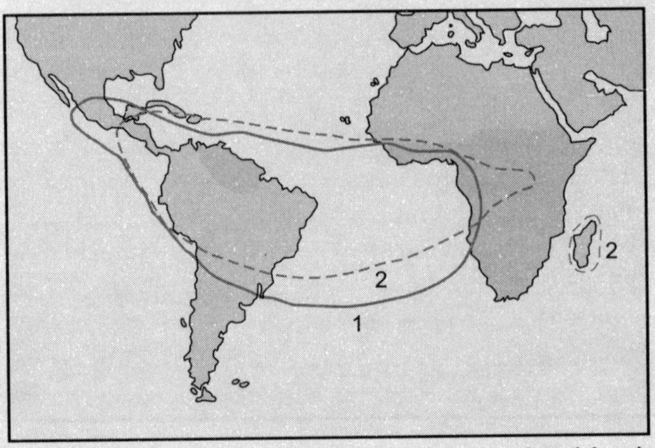

Figure 6.3. Concordant tracks of tropical trees (numbered 1 and 2) and tongueless frogs (light green), modified from Newbigin (1950: Figure 34).

Marion I. Newbigin was a celebrated Scottish geographer who received the Livingstone Medal from the Royal Scottish Geographical Society in 1924 for her contributions to the field of geography. Newbigin was well known for her influential books: *Animal Geography*, published in 1913, *A New Regional Geography of the World*, published in 1929, and *Plant and Animal Geography*, the first edition of which was published posthumously in 1936.

Newbigin's explanations for distribution patterns combined the mechanisms of dispersal and vicariance. The map, above, from the third edition of *Plant and Animal Geography*, published in 1950, depicts the concordant distributions or tracks of tropical trees (1, solid line *Vismia*; 2, dashed line, *Symphonia*) and tongueless frogs (horizontal lines, *Aglossa*). A mid-Tertiary European fossil closely related to *Aglossa* led Newbigin to propose a northern origin for the frogs. Yet she noted that these overlapping distributions "confirm the rather less definite evidence derived from the mammalian faunas that there has been in the geological past some kind of link between West Africa and Brazil despite the present width of the Atlantic here" (Newbigin, 1950:210).

cladogram in which the names of taxa have been replaced or augmented by the areas in which the taxa live can be interpreted in either one of two ways: as an areagram or as a taxon-area cladogram (TAC). In biogeography, areagrams are to TACs as cladograms are to phylogenetic trees in systematics. The area relationships as specified in one areagram can include those expressed in a variety of TACs, meaning that one areagram can be explained by a range of evolutionary processes. We restrict the definition of cladistic biogeography to conversion of taxon cladograms into areagrams and inference of biotic area homologs from such areagrams.[2]

Cladistic biogeographic methods are not phenetic nor are they matrix based. Cladistic biogeographic methods use topographical relationships among areas derived from taxonomic cladograms converted into areagrams. We outlined a set of principles for using areagrams to discover geographical congruence and biotic relationships and infer the changes among biotic areas through time in Chapter 4. Here, we describe two methods—component analysis, which incorporates Assumptions 1 and 2 of Nelson and Platnick (1981), and paralogy-free subtree analysis—which form the basis of subsequent methodological developments in cladistic biogeography that we explore more fully in Chapter 7.

Component Methods

Component analysis has its roots in the work of Platnick and Nelson (1978), who investigated the relationship between cladistics (phylogenetic pattern) and panbiogeography (geographic pattern). The premise is simple: if Earth and life evolve together, then relationships of taxa should reflect relationships of areas. Three taxa, A, B, and C, endemic to three areas, x, y, and z, respectively, are related as in Figure 6.4a. The area homolog is x(y,z) (Figure 6.4b). The area homolog can be explained by dispersal (from area x to y, and from area y to z) or vicariance (an ancestral distribution in xyz, with subsequent differentiation between areas x and yz, followed by differentiation between y and z). A *component* is a junction or terminal in an areagram that contains information about the relationships among the areas; the areagram of Figure 6.4a contains three terms, x, y, and z, and two components, yz and xyz, the latter trivial. Components alone convey no information about area homology, which is dependent upon topology.

In practice, analysis of area relationships is not as straightforward as in this hypothetical example: real biota usually present relationships more complex than three allopatric, endemic taxa in three areas (see Chapter 4). Areas repeat (are redundant or geographically paralogous) or may be missing (taxa may be absent from those areas or may be present but not collected). Taxa can be widespread (living in more than one area) and can form *MASTs* (Multiple Areas on a Single Terminal-branch)

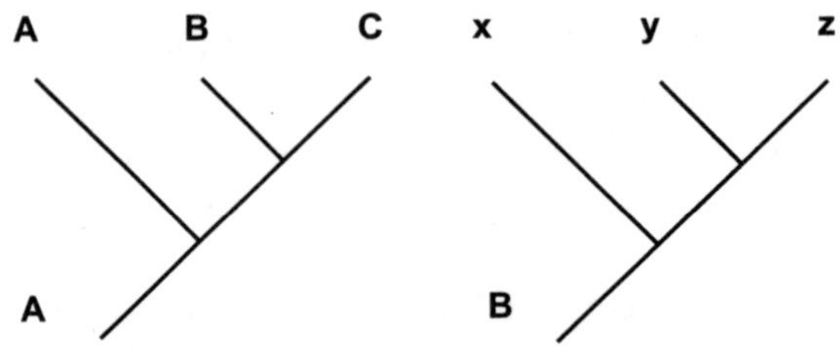

Figure 6.4. (A) Hypothetical cladogram representing relationships among three taxa, A, B, and C, that live in three areas, x, y, and z, respectively. (B) The areagram specifies the area homolog x(y,z).

in areagrams. These three phenomena—area paralogy, widespread taxa, and missing taxa—introduce ambiguity into areagrams that is addressed by different methods with different assumptions. Component methods are restricted to areagrams and have traditionally been done by hand, although several programs are available to implement them.[3]

Absence of taxa from a region requires no explanation. Yet missing taxa have been interpreted in biogeographic studies in various ways. The language a biogeographer chooses to describe a distribution often reveals a bias. To say that "cypriniform fishes have failed to reach South America" implies that dispersal is the critical mechanism of distribution. Such observations stifle biogeographic analyses, as they treat data on biotic diagnosis, homology, and relationships as irrelevant. Absence has been treated as a form of data that can be used with other information, such as that from sedimentology, geochemistry, and so on, to explain why taxa are absent. Ecological factors have been suggested to explain why many taxa broadly distributed throughout the Indo-Australian archipelago are absent from Borneo, for example. But the taxa (groups of plants, spiders, freshwater fishes, and so on) demonstrate the "entire range of ecologies" (see Heads, 2003:403–415). Different taxa can be absent from an area, such as Borneo, for different reasons—extinction, dispersal, failure to be collected, never having lived there, and so on. These factors can be considered to explore the history of individual portions of a biota after a general pattern has been identified.

Assumptions 1 and 2

Component analysis was developed by Nelson and Platnick (1981) to interpret the amount and kind of information about area relationships that could be extracted from widespread and redundant or paralogous areas on areagrams. They argued that these data could be interpreted using two,

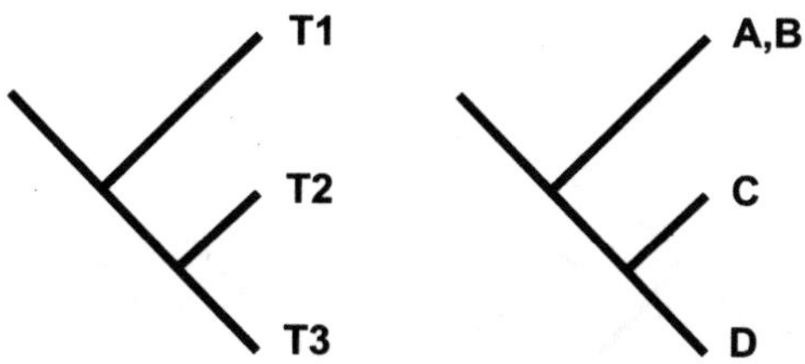

Assumption 1	Assumption 2	Assumption 0/BPA
A(B(CD))	A(B(CD))	(AB)(CD)
(AB)(CD)	(AB)(CD)	
B(A(CD))	B(A(CD))	
	A(D(CB))	
	A(C(DB))	
	B(D(AC))	
	B(C(AD))	

Figure 6.5. Possible relationships among four areas, A, B, C, D, occupied by three species, T1,T2, T3, under three assumptions, 1, 2, and 0, and Brooks parsimony analysis (BPA).

alternate assumptions: Assumption 1 and Assumption 2. Consider a hypothetical example (Figure 6.5): three species, T1, T2, and T3, live in four areas, A, B, C, and D. Species T1 lives in areas A and B; it is widespread, and its areas form a MAST. Species T2 lives in area C, and T3 lives in area D. What are the relationships of the areas as specified by the cladogram?

Under Assumption 1, if what we currently call species T1 will never be recognized as two distinct taxa, then what is true of one occurrence of the species T1 in area A is true of the occurrence of that species in area B. There are three possible dichotomous area relationships under this assumption: C and D are sister areas, and (1) A and B are sister areas, (2) A is more closely related to C and D than it is to B, and (3) B is more closely related to C and D than it is to A.

Under Assumption 2, if we allow that species T1 could be recognized as two distinct taxa in areas A and B, then what is true of one occurrence of the species T1 in area A may not be true of the occurrence of that species in area B. There are more possible relationships among the four areas under Assumption 2: all of those specified under Assumption 1, plus four more that allow areas A and B to be related to C and D (Figure 6.5).

A third assumption, Assumption 0, was proposed by Zandee and Roos (1987) as a biogeographic axiom that they felt Nelson and Platnick (1981) had failed to perceive. Under Assumption 0, areas occupied by

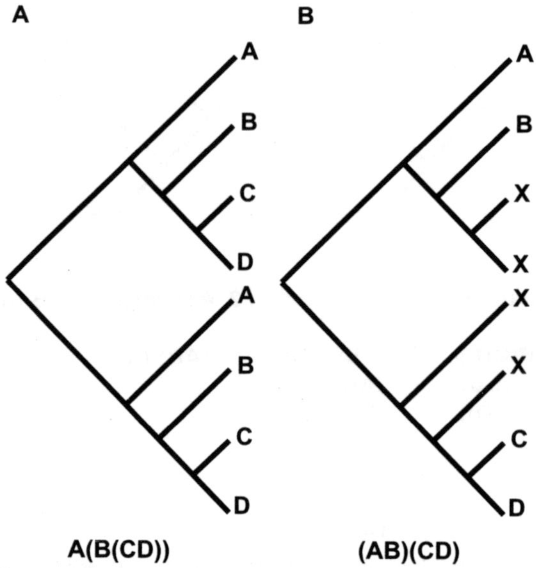

Figure 6.6. (A) Relationships among four areas, A, B, C, D, as specified by two duplicated lineages without extinction. (B) Extinction of four taxa, indicated by X, can lead to an erroneous conclusion of area relationships if the areagram is treated as a TAC.

widespread taxa must be closely related. This principle is also applied in Brooks parsimony analysis (BPA, see below). There is just one possible set of area relationships for the areagram under Assumption 0 and BPA: (AB)(CD).

Assumptions 0 and 1 conflict with the separation between phyletic and geographic components that is integral to Assumption 2. Assumptions 0 and 1 treat the areagram as a phylogenetic tree; Assumption 2 treats it as a summary of all possible area relationships. For example, if the "true" set of area relationships as specified by eight species is A(B(CD)) (Figure 6.6a), then extinction of several terminals (missing taxa) can lead us to conclude that area relationships are (AB)(CD) (Figure 6.6b). This means that nodes, such as (AB) in Figure 6.4b, hypothesized through a cladistic analysis of taxa, cannot be translated directly into geographic relationships or components. Repetition (lineage duplication) and extinction, both resulting in geographically paralogous nodes on areagrams, and widespread taxa or MASTs, characterize biotic evolution and must be accounted for in biogeography.

Paralogy-free Subtree Analysis and the Transparent Method

Paralogy-free subtree analysis was developed by Nelson and Ladiges (1996) to extract area homologs from areagrams by resolving geographic paralogy and MASTs. The method uncovers all the area relationships within an areagram and presents them in reduced areagrams that express the topographical relationships among the areas. For areagrams that contain multiple sets of relationships, more than one subtree can be recovered. We introduced this method in Chapter 4 and provide additional applications in Chapter 7; here we provide some historical context.

Although paralogy-free subtree analysis is straightforward, computationally it appears to be two methods: subtree analysis and what has been termed the *transparent method* (Ebach et al., 2005)—transparent, as opposed to "opaque to intuition" (Nelson's [1984:288] description of an application of Assumption 2). The transparent method clarifies how MASTs are resolved in a two-step process: (1) all possible area relationships as specified by the MASTs are depicted, and then (2) the paralogy-free subtree method is applied.[4] An example worked by Ebach et al. (2005) demonstrates the method:

Eleven "blue-ash" eucalypt taxa live in three areas of Australia, AU5 (southeastern forests), AU6 (Victoria and Tasmania), and AU7 (Adelaide) (see Ladiges et al., 1992). Their areagram includes two MASTs (underlined) and demonstrates rampant geographic paralogy (Figure 6.7a); it is likely a typical example of an areagram in a biogeographic analysis. To identify area homologs specified by this areagram, all possible area relationships are generated (Figure 6.7b). The two MASTs specify four areagrams, only one of which (outlined by a shaded box) is informative: AU7(AU6,AU5). Geographic paralogy of areas AU5 and AU6 adds no information on area relationship.

The result of a paralogy-free subtree analysis—a general areagram—is free of MASTs and geographically paralogous areas and represents all that we know about the relationships of areas given the data at hand. It is a hypothesis of area relationships that can be tested by the analysis of additional taxa in the areas under study.

EVOLUTIONARY BIOGEOGRAPHIC METHODS

Many methods used in biogeography are classed as evolutionary biogeographic methods. They apply techniques from ecology, paleontology, or molecular systematics, for example, to predict or postulate

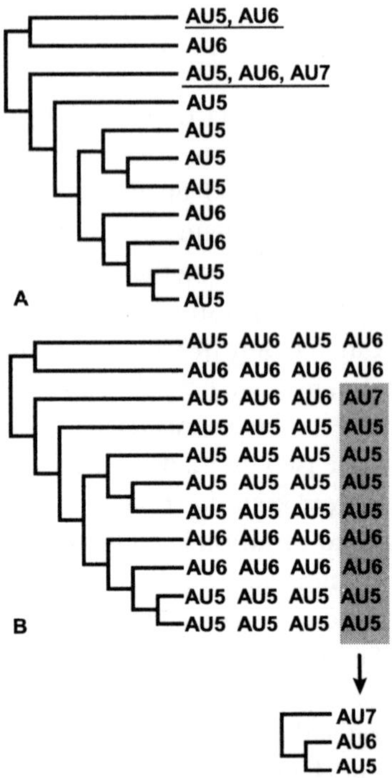

Figure 6.7. (A) Areagram of 11 taxa of "blue-ash" eucalypts that live in three areas, AU5, AU6, and AU7. (B) Area relationships are resolved by generating all possible area relationships, and then identifying those that specify an area homolog: AU7(AU6,AU5). (Modified from Ebach et al., 2005.)

evolutionary and/or geographical mechanisms to explain distribution patterns. Evolutionary biogeographic methods focus on generating hypotheses for distributions based on evolutionary, ecological, and geographical explanatory mechanisms, often outside of a phylogenetic framework (e.g., Ricklefs, 2004).

Ecographic Methods

Ecography is the study of spatial ecology with the goal of explaining ecological distribution patterns. The range of ecographic methods is vast, and a detailed review of the field is beyond the scope of this book. We address just one set of methods that has a strong mapping—hence, biogeographic—component: niche modeling.

Niche Modeling Data used to generate taxon distribution maps—such as those from museum or herbarium specimens—may be highly accurate with respect to species identification, but may be only an incomplete, biased reflection of natural distributions. The distribution map of pipefishes (Figure 6.1a), for example, although accurate, is not precise: pipefishes do not live throughout the Indonesian island of Sumatra even though it is blackened on the map, and they also do not live continuously throughout the shaded areas of the Pacific Ocean.

The development of more sophisticated mapping techniques, such as GIS (Box 6.1), which pinpoint location of specimens using georeferenced data, and the modeling of species ecological attributes, such as temperature, depth, elevation, and so on, have led to the development of *niche modeling*. Ecological attributes of live species can be used to model their habitats, to more accurately map their distributions, and to more accurately predict where they live. Subtle changes in salinity, temperature, or depth may mark the limits of a pipefish species distribution, and such changes are likely to go undetected without access to GIS-based environmental characteristics.

Methods to predict species distributions from collection data are being developed and tested (e.g., Elith et al., 2006), while databases of ecological attributes are being updated and expanded. The techniques have already led to numerous applications in conservation biology and can be expected to aid biogeography in many ways, not the least of which is in more accurately describing distributions.

Phylogenetic Methods

Phylogenetic methods use phylogenies and evolutionary and/or ecological mechanisms to explain taxic or biotic changes through time. Phylogenetic methods may use TACs or inferred genealogies. The majority of biogeographic methods are phylogenetic or chorological, as they are reliant on ancestor-descendant concepts (e.g., transformation of characters, taxa, or hypothetical ancestors).

Phylogenetic biogeography originated in the work of Ernst Haeckel (1866), who used a phylogenetic tree, instead of a biostratigraphic sequence, to postulate distributional pathways and centers of origin. The paleobiogeographical/neobiogeographical division reinforced by the use of biostratigraphy versus phylogeny is arbitrary and reflects the historical division between those who do paleontology and those who do not (see Chapter 7).

The modern application of phylogenetic biogeography was pioneered by entomologists Willi Hennig (1966) and Lars Brundin (1966). The inferred most primitive taxon in a cladogram or TAC was assumed most likely to be located in the center or origin or in part of the ancestral area. The progression from the most primitive node to the most derived was interpreted as the direction of dispersal, a concept that became known as the Progression Rule. Critical to the development of phylogenetic methods in biogeography was the interpretation of the distribution of one taxon at a time: the Progression Rule is not comparative. To apply the Progression Rule is to incorporate an explanatory mechanism into a biogeographic method (see Chapter 5 for a further discussion).

Phylogeography sensu Avise (2000) and Avise et al. (1987) is a collection of methods that use molecular data to hypothesize genealogical pathways among populations or individuals within a species to infer the process by which they became geographically distributed (see Riddle, 2005). Phylogeographers use known ecological, phylogenetic, and geographical mechanisms to explain the distribution of a single phylogeny. Questions asked in a phylogeographic study include, Is degree of genetic differentiation correlated with geographic distance separating populations? When framed in that way, hypotheses are testable. Numerous phylogeographic studies invoke the Progression Rule without criticism and infer migration of populations away from a center of origin as a distributional mechanism. Applications of *comparative phylogeography,* the comparison of patterns among species, have relied on existing tools for comparison of molecular phylogenies; among the most popular tools is BPA (Lomolino et al., 2005; see below).

Phylogenetic Analysis for Comparing Trees (PACT) is a method proposed by Wojcicki and Brooks (2005) for comparing and combining TACs generated from phylogenetic analyses. Describing the method is a good way to demonstrate the difference between the goals of phylogenetic and cladistic biogeography, as defined above.

PACT assumes that TACs are like trees and not like cladograms. PACT summarizes the area relationships for a group of taxa by combining TACs. A PACT-derived areagram for five areas, A, B, C, D, and E, as specified by a group of TACs, is A((BE)(C(DA))) (Wojcicki and Brooks, 2005: Figure 9).

Area A is geographically paralogous on the summary TAC. Its relationship to the other areas is ambiguous. In contrast, Wojcicki and Brooks (2005) argue that this is the ideal way to demonstrate the relationships of hybrid or composite areas: A is related to D and also to areas BCDE. The statement C(DA) is an area homology. The statement A(BCDEA) conveys contradictory information about area relationships and contradicts the explicit area homolog, C(DA). Contradictory information on area relationships, such as that from composite areas, may be expressed at different stages of the hierarchy (see Chapters 7 and 9) and does, of necessity, require reticulations or polytomies, which convey no particular area relationship (see Chapter 4).

Matrix-Based Methods

As methods that incorporated phylogeny into biogeography were developed, it was proposed that just as characters could be used to infer relationships among taxa, taxa could be used to infer relationships among areas. Systematists built a taxon-by-character matrix; biogeographers built an area-by-taxon matrix (e.g., Crovello, 1981). How to build and analyze such matrices has been the focus of much biogeographic debate and has stimulated the development of an array of methods. Matrix-based methods are popular because of their versatility; they identify a diverse array of biogeographic patterns. Matrix-based methods use dendrograms, phenograms, cladograms, or no phylogenetic hypothesis at all. Any phylogenetics software can implement a matrix-based method and quickly produce an area hierarchy. Matrix-based methods differ from cladistic methods, as we define those above, in a critical way: matrix-based methods do not identify area homologs. We review three groups of matrix-based methods here: parsimony analysis of endemicity (PAE), Brooks parsimony analysis (BPA), and ancestral area analysis (AAA), and their variants, which span the array.

Parsimony Analysis of Endemicity (PAE) was developed by paleontologist Brian R. Rosen (1985, 1988) to uncover the similarity of biota in the absence of phylogenetic data. PAE was devised as a way to apply a pseudo-cladistic analysis, including applying parsimony, when the only data available were the presence or absence of taxa in areas.[5]

	A	B	C	D	E
T1	0	0	0	1	1
T2	0	0	1	1	1
T3	0	1	1	1	1
T4	1	1	1	1	1

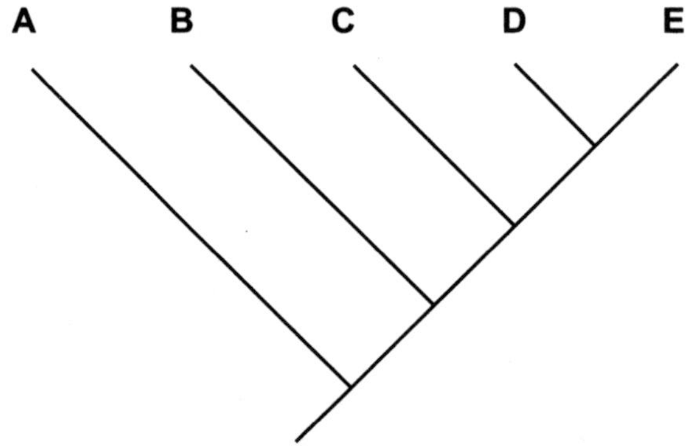

Figure 6.8. Parsimony analysis of endemicity (PAE) is a matrix-based, phenetic method. A taxon-by-area presence/absence data matrix is constructed, and the matrix (above) is analyzed using a parsimony algorithm or by hand. Four species (T1, T2, T3, T4) live in five areas (A through E), which cluster as in the area phenogram (below).

Each area is assigned a value of 0 (absence) or 1 (presence) based on the taxa that live there. Consider the taxon-by-area presence/absence data matrix for five hypothetical species that live in areas A, B, C, D, and E (Figure 6.8). Species T1 lives in areas D and E; those areas are given a value of 1 in the data matrix. Likewise, species T2 lives in areas C, D, and E; species T3 lives in areas B, C, D, and E; and species T4 lives in all five areas.

The most parsimonious distribution of the species among the areas of endemism is calculated by hand or using a parsimony program using an all-zero outgroup: absence is primitive and presence is derived. Other rooting methods can be used; the biogeographic interpretation of any rooting method for a PAE-derived network is debatable (Rosen and Smith, 1988). PAE does not aim to discover and test for area homologs. PAE is a way to generate hypotheses based purely on the similarities of areas. As implemented by paleontologists Rosen and Smith (1988),

PAE was used to demonstrate changes in species composition of areas through time.

The result obtained by PAE is not an areagram or a TAC, in the sense that we use those terms, but an *area phenogram.* PAE conveys no phylogenetic information about areas or taxa.

Cladistic Analysis of Distributions and Endemism (CADE) (Porzecanski and Cracraft, 2005) is a matrix-based method developed by ornithologist Joel Cracraft (1991) to interpret historical relationships among areas of endemism when no phylogenetic relationships are available. Like PAE, it builds a presence/absence by area matrix that is analyzed using a parsimony algorithm. As implemented by Porzecanski and Cracraft (2005) for South American birds, it differs from PAE in several ways. Areas are not simply locality points, but areas of endemism, as determined and agreed upon in a range of previous studies. Hierarchical relationships of taxa (e.g., species and genera) are used as a proxy for phylogenetic relationship. And large datasets are analyzed. The differences in these methods have been characterized as primary PAE (e.g., as in Rosen, 1988) and secondary PAE (as in CADE; see Nihei, 2006).

CADE is a method of area similarity, not area homology. It is a way to summarize available information on distribution *absent* phylogenetic relationships. Hierarchical relationships among areas as generated by CADE can be tested by phylogenetic analyses of taxa, which themselves may be used to generate another set of hierarchical area relationships or area homologs.

Nested Areas of Endemism Analysis (NAEA) is another method that implements a CADE or secondary PAE-like analysis (Deo and DeSalle, 2006). NAEA is a derivative of nested clade analysis (NCA), developed by Alan Templeton (1998) to interpret phylogeographic data. NCA starts with a cladogram or network and proposes mechanisms by comparing genetic similarity with geographic distance (as in Sokal, 1979). NAEA implements NCA and PAE together to find similarities between areas based on nested sets within TACs. NAEA assumes that areas that share more species are more closely related to each other than they are to other areas. Furthermore, areas are grouped together if they share similar species distributions. Hierarchical relationships of species assemblages as generated by an overall similarity method such as NAEA, as with CADE, need to be tested by area homologs as generated from phylogenetic analyses of taxa in the areas of interest.

Brooks Parsimony Analysis (BPA) (Wiley, 1987) was derived from the applications of cladistic methodology to the study of host/parasite relationships by parasitologist Daniel Brooks (1981, 1985). The method assumes that being infested with one species of parasite is evidence for the monophyly of a group of hosts, and that if one host is infested with more than one species of parasite, then that group of parasites is monophyletic. The analogy between host/parasite relationships was extended to taxon/area relationships in biogeographic applications of BPA.

Unlike the aforementioned matrix-based methods, BPA requires a phylogenetic hypothesis in the form of a cladogram as a starting point. Names of taxa are replaced by the areas in which they live to create a TAC, not an areagram. All terminals and nodes are numbered or named, and an area-by-terminal/node matrix is built. Missing areas are coded as "unknown." The same species living in more than one area, a MAST, is coded as a "synapomorphy." This was described above as Assumption 0 (Figure 6.5).

Ways to implement BPA in biogeography and the study of host-parasite relationships have been discussed and debated extensively since the method was first proposed (e.g., Brooks et al., 2001; Lieberman, 2000; Siddall and Perkins, 2003; Brooks et al., 2004; Siddall, 2004). The original implementation has been termed primary BPA. In secondary BPA (Brooks et al., 2001) areas that are inferred to have a reticulate history are duplicated on the summary TAC (see also PACT, above). We acknowledge the debate and argue that when TACs or areagrams are reduced to binary data to build secondary TACs, information on area relationships is obscured or lost (compare Assumptions 2 and 0 in Figure 6.5; see also Ebach et al., 2003).

Ancestral Area Analysis (AAA) was devised by botanist Kåre Bremer (1992, 1995) to estimate the ancestral area of a taxon by interpreting the relationships among areas as specified by a TAC. It assumes a center of origin for a group with subsequent dispersal and differentiation as taxa moved away from that center or ancestral area, the ancestral distribution of that group. As in BPA, a TAC, not an areagram, is assumed, and the relationships among areas are hypothesized from one TAC. In AAA, the gains (presence) and losses (absence) of taxa in areas on the TAC are tallied. The function losses/gains provides a value (the higher the value the more likely it is part of the ancestral area). In the simplest TACs, a high number of gains to losses specifies an ancestral area. In more complex TACs, assumptions about the likelihood of gains versus losses must be made to interpret the ancestral area. A taxon-by-area matrix may be built and analyzed through a parsimony program to infer a cladogram on which areas are optimized. AAA treats all areagrams as TACs and requires optimization of areas. We have discussed above (and in Chapters 4 and 5) why we reject this approach. Chief among those reasons is that duplicated areas are more likely to be interpreted as part of the ancestral area than are single areas; thus, ancestral areas are interpreted based on where species happen to survive.

Dispersal and Vicariance Analysis (DIVA), devised by Fred Ronquist (1997, 1998), is a modification of AAA in which the events or mechanisms of extinction, dispersal, vicariance, and sympatry (lineage duplication) are associated with a cost. These values represent hypothetical events within a given model (e.g., ancestral areas, vicariance, dispersal, etc.) and are constructed as a value-by-area data matrix that is analyzed using parsimony, maximum likelihood, or Bayesian algorithms. The resultant TAC is a hypothetical model of events that may be interpreted as an evolutionary and biogeographic history of the group in question. The method interprets how well a particular TAC can be explained by a series of events. A given series of events may have had nothing to do with the evolution and distribution of the group, however. To reject vicariance in favor of dispersal as an explanation for differentiation of two lineages because their time of differentiation, estimated using a molecular clock, is inferred to be later than that of a particular vicariance event is trivial. Other vicariance events may be responsible for the differentiation, dates may be over- or underestimated (for both the lineage differentiation and the vicariance event), and so on (see below).

Other variants of AAA that use a value system of gains and losses or that assign values for particular biogeographic events are weighted ancestral area analysis (Hausdorf, 1998) and resolved areagrams (Enghoff, 1996).

BIOGEOGRAPHIC APPLICATIONS

Many applications thought of as unique to biogeography have been developed in other fields such as co-evolution, geography, paleontology, ecology, and so on. Here, we classify biogeographic applications as systematic (those concerned primarily with topographical relationships of areas) or evolutionary (those concerned primarily with mechanisms). Applications used in systematic biogeography are mostly topographic and aim to uncover additional biogeographic patterns or congruence. The majority of biogeographic applications is evolutionary and aims to uncover the mechanism of distribution.

Systematic Biogeographic Applications

Geographical Techniques

Geographical techniques use geographical data and/or geographical congruence to propose the former positions of biota and to infer biotic limits. They require either general areagrams or hypothetical geological areagrams that can be drawn as topographical maps.

Terrane Analysis, a robust method for generating hypothetical area classifications in the form of areagrams was developed by paleontologist Gavin Young (1984, 1995) and is designed to accommodate further systematic biogeography analysis. Terranes are fault-bounded, geological formations. They may remain separate or become accreted to continents. Terranes are extremely valuable in determining the shape and extent of geologic and biogeographic areas over time. "Terranes" may also be interpreted as ecological zones, reef systems (areas of bioaccumulation), ecological succession, and so on.

Terrane analysis operates from the premise that areas, whether geological, ecological, or biotic, are dynamic and change over time. Terrane movement or accretion is represented hierarchically in a branching diagram. Terrane fragmentation is analogous to lineage splitting; terrane accretion is analogous to coalescence of lineages. One aim of terrane analysis is to retrodict how terranes may have been positioned geographically. The proximal positions can be drawn as an area phenogram that equates similarity with distance. The area phenogram is a hypothetical classification, one that retrodicts the relationships of biota during the time of biotic divergence (geographical isolation). Area phenograms can also be drawn as geographical reconstructions on maps. Although this is far more practical for paleontological data, it can also be effectively used with extant biotic areas.

When proposing a hypothetical classification of areas for a systematic biogeographic analysis, the past boundaries of an area should be considered. Over time, some ecological zones may have shifted: the meandering of a river through different soil and rock types creates different types of ecosystems. A river running through clay to loamy soil may form swamps and lakes. Rivers that meet a transgressing shoreline may form river-dominated deltas, such as the Mississippi River Delta. Ecological zones that change over time may form different proximal relationships to other areas. Oceanic islands also change position and form different relationships to each other and to continental landmasses. Terrane analysis was developed to classify these changes in areas over time.

Development of the west coast of North America is among the best examples of terrane accretion. Oceanic or island arc terranes were accreted onto the west coast of the North American continent during the Late Triassic to Middle Cretaceous geological periods. The restricted distributions of both living and fossil taxa have been used to identify the terranes. Bolitoglossine salamanders, for example, demonstrate "terrane fidelity" or terrane endemism throughout the western United States (Hendrickson, 1986).

Panbiogeography, in its attempt to unify geological and biological data, extended assumptions about biology to geology. The assumptions about phylogenetic relationships of taxa have not been applied directly to areas. The approaches of paleobiogeographers such as Young (1984) and Rosen and Smith (1988) hypothesize area relationships through time without a systematic analysis of geological/geographical characteristics that could be the "synapomorphies" of geological cladograms. The shortcomings of such analyses come more from lack of identifiable geological/geographical homologies than from faulty methodology.

Identification of geological "homologies" to form geological cladograms that could be compared with biological areagrams was an unrealized goal of Donn Rosen (see Parenti, 2006). Geological and geographical evidence may be ambiguous, as similar terranes in the same distributional areas may not necessarily contain the same types of biota. The same holds for paleobiogeographical regions. One could predict that a shallow marine sedimentary basin in a former epicontinental sea during the Ordovician *should* include trilobites, but it nevertheless may not.

Area Cladistics, a technique developed by Malte Ebach (1999) and colleagues (Ebach and Edgecombe, 2001; Ebach and Humphries, 2002), builds on Young's terrane analysis. Terrane analysis and area cladistics are complementary.

Area cladistics interprets area relationships as *geographical distance.* The component in a general areagram is assumed to indicate biotic divergence (i.e., geographical isolation). The processes assumed during biotic divergence are *descriptive vicariance* and *descriptive dispersal*; no other explanatory mechanism is inferred. The underlying assumption of area cladistics is that general patterns result from the biotic relationships and position of former areas. If we were to examine tropical South American and African biotas, currently separated by the Atlantic Ocean, we would find that many components of the biota share a closer relationship to each other than either does to other areas—even to geographically closer areas, such as Arabia relative to Africa.

Area cladistics requires geographical congruence as specified by more than one areagram. Single areagrams only show parts of patterns, and their components indicate nothing more than a group of areas.

Biogeographic reconstruction is a way to express geographical congruence or other biogeographical hypotheses. Such reconstructions are used to infer the topographical, geological, geographical, or ecological separations or discontinuities that are hypothesized as mechanisms for past or current distributions, as well as to infer past pathways for dispersal or migration. Biogeographic reconstructions are made with geographical or geological maps; biogeographers may superimpose past distributions over paleomaps based on geological reconstructions. Polychaetes are an ancient lineage of marine worms with a pelagic larval stage; their biogeography has traditionally been interpreted with respect to long-distance dispersal throughout the global marine realm (e.g., Fauchald, 1984). An areagram for polychaetes was generated using BPA and then interpreted with respect to currently accepted geological reconstructions by Glasby (2005; Figure 6.9). Polychaetes mirror the distribution of other taxa, both marine and terrestrial, in having boreal, austral, and pantropical distributions—global realms recognized at least since the mid-19th century (see Chapter 2). The areas, or global realms, were related in the area homology Austral (Boreal, Pantropical) using BPA. Glasby concluded that this distribution pattern could be explained by a combination of vicariance and biotic dispersal or dispersion.

The reverse—geological reconstructions based on area relationships as inferred from taxa—are rarer (see Ebach and Edgecombe, 1999; Ebach and Humphries, 2002). These are made separately from paleomaps, or with paleomagnetic, rather than with traditional, geological data (i.e., paleogeography, see below). Biogeographic maps constructed

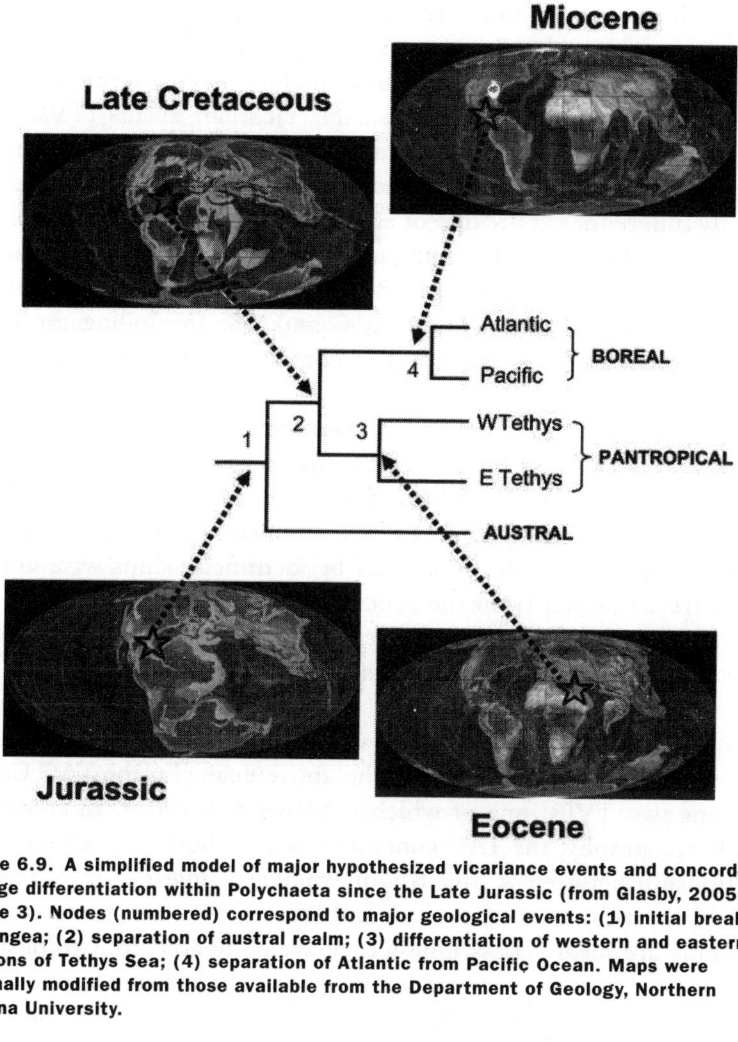

Figure 6.9. A simplified model of major hypothesized vicariance events and concordant lineage differentiation within Polychaeta since the Late Jurassic (from Glasby, 2005: Figure 3). Nodes (numbered) correspond to major geological events: (1) initial break-up of Pangea; (2) separation of austral realm; (3) differentiation of western and eastern portions of Tethys Sea; (4) separation of Atlantic from Pacific Ocean. Maps were originally modified from those available from the Department of Geology, Northern Arizona University.

in this way provide independent evidence for paleogeography and may be used to test paleoreconstructions where geological data (e.g., paleomagnetics) are absent. Biogeographers can inform geologists of the past configuration of continents or oceans (see Escalante et al., 2007; McCarthy et al., 2007) based on empirical data derived from phylogenies (i.e., area cladistics; Ebach, 2003).

Hovenkamp Vicariance Analysis (HVA) (Fattorini, 2008), includes several methods named after their developer, Peter Hovenkamp (1997,

2001). These methods aim to reconstruct Earth history as a sequence of vicariance events rather than focusing on relationships among areas.

The first variation of HVA (Hovenkamp, 1997), termed HVA1, begins with a TAC. So-called traceable vicariant events (TVEs) are identified for nodes that contain allopatric sister taxa; the argument is that such sister taxa logically may have become disjunct and subsequently differentiated because of a vicariant event. The method allows that a single TVE could be explained by the mechanisms of vicariance or dispersal. A TAC, such as C(AB), contains two TVEs: one between A and B, and the second between C and AB. Compiling the same TVEs in various TACs is treated as corroborating supported vicariance events (SVEs).

The second variation of HVA (Hovenkamp, 2001), termed HVA2, also begins with a TAC, which is mapped to examine visually the distribution of taxa in nodes. Maps that demonstrate allopatric distributions of taxa in nodes, or TVEs, are retained. Maps with TVEs are grouped together so that SVEs may be identified. Maps with unique TVEs are discarded from the process, but they may be retained for further analyses. SVEs may be arranged linearly or in a reticulate pattern to demonstrate the Earth history events that formed the distribution.

Both implementations of HVA are incompatible with our method of comparative biogeography. In the above example, the TAC C(AB) contains two TVEs, one of which is between A and B. In comparative biogeography, the TAC contains one area homolog: C(AB). The relationship between A and B is meaningless unless it is compared to C. Even if the taxa of areas A and B are sister taxa, the area relationships are meaningless unless they are compared to a third area. To assume otherwise is to assume that the split between areas A and B must have been the mechanism by which the taxa in those areas became differentiated.

Furthermore, HVA dismisses the power of areagrams in discovering new geological reconstructions that challenge geological orthodoxy: "HVA1 and HVA2 appear to be empirically superior to all other methods not because some of their results are compatible with those produced by other methods, but because their results are in good accordance with our present knowledge of earth history" (Fattorini, 2008:620). Agreement of species distributions with known geological events may be satisfying, but it does not allow the discovery of new geological arrangements. The method agrees with our current knowledge of geology because it is dependent upon it. Ultimately, implementations of HVA, as detailed in Fattorini (2008), do not produce testable hypotheses that are independent of our current knowledge of Earth history.

Methods that produce reticulations, of area relationships or SVEs, have been deemed superior to resolved areagrams (see especially Fattorini, 2008) because they are thought to be more realistic: Earth history is reticulate; therefore, biogeographic patterns should be reticulate too. As in our discussions throughout this book, we view this as a failure to recognize the difference between areagrams and TACs, poor area delineation, or a combination of the two. Reticulations expressed as polytomies convey no information about area relationships. Reticulations that depict an area as being in more than one set of sister-area pairs do not propose testable area homologs.

Evolutionary Biogeographic Applications

Geological Techniques

Geological techniques use geological data (e.g., petrological, geochemical, paleomagnetic, sedimentary, and fossil data) to classify former terranes and reconstruct the former biota and geographical positions of accreted terranes without using areagrams or TACs. These techniques rely on geological and stratigraphical records as well as on paleoecological and paleoclimatological data derived from paleomagnetics, geochemistry, or sedimentology.

Paleogeography aims to reconstruct the extent and positions of continents and oceans through time. Paleogeography proposes geological reconstructions based on certain types of geological data (i.e., paleomagnetics, paleoecology, etc.). All paleogeographic data are extracted from the geological or climatic record (e.g., from ice cores or through dendrochronology). Paleomagnetic data are among the most widely used and robust kinds of paleogeographic data. Rocks, such as basalts, which contain magnetic minerals that have formed quickly, can preserve information about the position of Earth's magnetic field. Paleomagnetics measures the alignment of magnetic minerals in older rocks to calculate the positions of continents during mineralization. Most continents have undergone tectonic movement, and the rocks that form them have continually been reheated and deformed. The older the rock, the more likely it has been reworked and reheated, meaning that most, if not all, of the original magnetic alignments have been reset.

Another problem of paleomagnetics is that it can only tell us about paleolatitude (Kearey and Vine, 1996). Paleolongitude is calculated by a series of different and inconsistent methods that use sedimentology, or the occurrence of fossils, as general indicators. Despite these drawbacks, paleogeography has been used to corroborate biogeographic patterns, especially in terrane analysis and area cladistics.

Other paleomagnetic techniques include geochemical (isotope) analysis to retrodict the composition of atmospheres and oceans (Kearey and Vine, 1996) or to study reef building organisms that reveal sea level rises and falls (transgressions and regressions) or changing geology and geography of island arcs, the movement of which opens and closes seaways (Metcalfe, 2001).

Temporal Techniques

Biogeography and systematics have adopted an array of methods for dating a node on a phenogram or cladogram by comparing it with a fossil of known age. With fossils as their basis of calibration, molecular evolutionary clocks use changing rates in molecular mutations to date nodes. Recently, there has been a revival in estimating the absolute ages of nodes based on fossils or molecular clocks. Although molecular clocks are non-biogeographic, they have been used to hypothesize different biogeographic patterns. Many argue that absolute ages of taxa should be assessed prior to a biogeographic analysis in order to compare patterns of taxa from the inferred same geological period (e.g., Donoghue and Moore, 2003). Thus, absolute ages of taxa are used to generate biogeographic patterns. Our view, in contrast, is that the hypothesized age of a taxon should be part of the *interpretation* of biogeographic patterns that are based on the analysis of area relationships, not the *generation* of biogeographic patterns (Ebach and Humphries, 2002). The age of a taxon can best be estimated from the biogeographic pattern. Using the inferred absolute age of a taxon (or molecular, fossil, or stratigraphic data) constrains or restricts the variety of possible area relationships that might be revealed. Temporal techniques differ based on whether one seeks the ages of the terminals or nodes or whether one attempts to date geographical areas or biota.

Stratophenetics and Stratocladistics aim to assign either absolute or relative ages to lineages based on the absolute or relative ages of fossil taxa. Stratophenetics (Gingerich, 1979) was developed as a method to estimate the evolutionary sequence of fossils with their relative placement on a timeline in the fossil record. Temporal data joined morphological and distributional data to estimate an evolutionary sequence drawn as a phenogram. Stratocladistics (Fisher, 1994, 2008; Fox et al., 1999) uses fossil occurrence to interpret and generate cladistic phylogenetic hypotheses. Neither method was developed strictly for biogeography, but fossil occurrence implies geography, as well as some estimate of age.

A problem of stratigraphically focused methods is that they assume that the oldest fossil represents the first occurrence of a lineage and that the fossil is the ancestor (real or hypothetical) of a lineage. It has been argued that fossils represent only the minimum ages of taxa, yet these methods counter with the assumption that despite the failings of the fossil record, it can be used to hypothesize evolutionary and biogeographic history. Stratophenetics and stratocladistics remain tautological: it is impossible to *confirm* that the oldest recorded fossil is both the *oldest* taxon and the *ancestor* of all other taxa within a lineage. Cladistics rejected stratophenetics on the grounds that it bases age on an exact timeline, whereas relationship in each cladogram and phenogram was either proximal (relative) or based on a probability, respectively. Chronobiogeography (Hunn and Upchurch, 2001) is also dependent on a fossil or ancestral age to date nodes.

Temporal Geographic Paralogy aims to place exact ages onto the nodes of cladograms have often been misdirected. Cladograms have been matched to stratigraphic columns to add the time dimension to phylogenies. The problem of dating nodes in cladograms or areagrams becomes apparent when two clades have different timeframes. Zaragüeta et al. (2004) demonstrated that ages can be depicted relatively using temporal geographic paralogy. The technique simply dates the nodes in cladograms based on the ages of the terminals. Nodes on the cladogram are dated relatively (i.e., the oldest members of a clade tend to be interpreted as "basal" [e.g., Janvier, 2007; but see Krell and Cranston, 2004]). The underpinning principle of temporal geographic paralogy is that hierarchical sets "relate" relative periods of time. The Cretaceous is younger than the Jurassic, the Jurassic is younger than the Triassic, and so on. The set can be expressed as (Triassic (Jurassic (Cretaceous, Cretaceous))) without compromising a chronological hierarchy. The temporal paralogy can be applied to cladograms as well as areagrams to classify clades into time periods. The pectinate (comb-like) tree (Figure 6.10a) summarizes the relationships among six taxa (A through F) and their ages (time periods 6 through 1, from the oldest to the most recent). Correspondence between age and position in the hierarchy is precise. Finding a new fossil, N, of age 4 (Figure 6.10b), erodes the correspondence between age and hierarchical position, as now two different timeframes are represented. The sister clades (CN) and (DEF) must be inferred to be of the same age; in other words, they must be temporal paralogs (Figure 6.10b).

Basing the age of nodes or components on the ages of the biotic areas or taxa at the terminals is a less ambiguous approach to dating areagrams than is using hypotheses of age based on fossils or molecular clocks. Temporal geographic paralogy also avoids "splitting" areagrams into different ages. The important factor in dating areagrams is the age of the biotic areas, not the lineages or nodes. Area delineation is known prior to an analysis; therefore, two geographical areas (as in the case of the rockwarbler and the extinct amphibian labyrinthodont from the Sydney Basin; see Chapter 4) can share *two* different relationships because they are *two* separate biotic areas.

We have discussed a range of methods that have been applied to discover and interpret biogeographic patterns and the mechanisms that may have formed them. Some methods do not attempt to discover a pattern and, instead, interpret organismal distribution using mechanisms alone. Other methods interpret biogeographic patterns with respect to an independent geological history. We aim to unite these approaches: to discover biogeographic congruence and interpret it relative to a range of possible mechanisms, both biological and geological. In Chapter 7, we develop more fully the method that we introduced in Chapter 4. We apply these principles to geological and biological evolution of the Pacific in Chapter 9.

SUMMARY

- Methods and applications used in biogeography can be categorized as systematic or evolutionary.
- Many methods and applications are based on available techniques of ecology (i.e., island biogeography), molecular

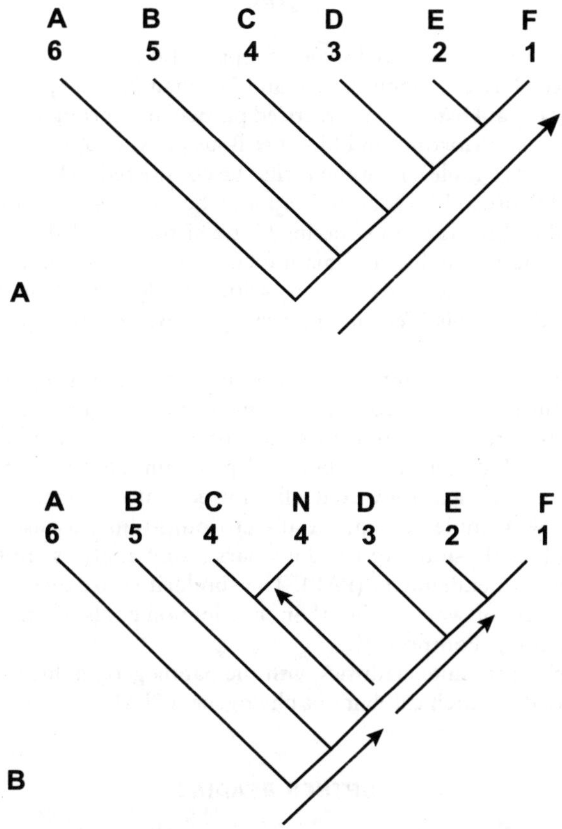

Figure 6.10. An example of temporal information contained in cladograms (modified from Zaragüeta et al., 2004: Figure 3). The arrow is time. (A) Taxa (A through F) and their ages (time periods 6 through 1) correspond with time's arrow. (B) Discovery of a new fossil, N, at time 4, creates a paralogous timeframe for sister taxa (CN) and (DEF), which logically are of the same age. See text for further discussion.

systematics (i.e., phylogeography), phenetics (i.e., PAE), phylogenetics (i.e., BPA), cladistics (i.e., paralogy-free subtree analysis), and paleontology (terrane analysis, stratophenetics, stratocladistics).

- Cladistic biogeography is restricted to component analysis and paralogy-free subtree analysis and differs from other methods of area classification that include matrix-based methods because it recognizes area homologs.

NOTES

1. Panbiogeography received strong support also in New York, through the work of Gareth Nelson, Donn Rosen, and Norman Platnick, at the American Museum of Natural History. It is practiced particularly throughout Mexico and South America (see Morrone and Llorente-Bousquets, 2003).

2. Crisci (2001: Table 1) outlined what he considered to be main characteristics of nine historical biogeographic approaches. Just one, panbiogeography, was described by him as reconstructing biotic history. Cladistic biogeography, as we define it here, also has as a main focus the history of biota.

3. COMPONENT 2.0, a program written by Roderic Page (1993), may be downloaded from his Web site: http://taxonomy.zoology.gla.ac.uk/rod/cpw.html.

4. Prior to the development of non-matrix-based programs such as *Nelson05* (Ducasse et al., 2007), implementation of subtree analysis was restricted to programs that relied on current and readily available phylogenetics software. TASS, an MSDOS-based program, converted an input tree into subtrees, and then converted all subtrees into a matrix. As no area optimization was implemented, an all-zero outgroup was used. The TASS matrix resembles those used in Brooks parsimony analysis (BPA) or phylogenetic analysis of endemicity (PAE). The similarities are due to the reliance on phylogenetics software, rather than on common goals of the methods (see Ebach et al., 2005: Footnote 1).

5. Attempting to unite cladistics with the panbiogeographic method, Craw (1988:295) used hierarchical, but not phylogenetic, PAE.

FURTHER READING

Forey, P. L. 1982. Neontological analysis versus palaeontological stories. In K. A. Joysey and A. E. Friday (eds), *Problems of phylogenetic reconstruction* (pp. 197–234). Academic Press, London.

Humphries, C. J. 2004. From dispersal to geographic congruence: Comments on cladistic biogeography in the twentieth century. In D. M. Williams and P. L. Forey (eds.), *Milestones in systematics* (pp. 225–260). CRC Press, Boca Raton, Florida.

Lomolino, M. V., D. F. Sax, and J. H. Brown (eds). 2004. *Foundations of biogeography: Classic papers with commentaries.* University of Chicago Press, Chicago.

Page, R. D. M. 1990. Component analysis: A valiant failure? *Cladistics,* 6, 119–136.

SEVEN

THE SYSTEMATIC BIOGEOGRAPHIC METHOD

DOING SYSTEMATIC BIOGEOGRAPHY

Hypotheses of area relationship for the monophyletic taxa of a biota are vital for constructing a biogeographical classification, and thus vital for uncovering the ontological divisions of the Earth. Any technique that

Overview

The goal of systematic biogeographical analysis is to find congruence between two or more area relationships: geographical congruence or *area monophyly.* We explain the importance of area relationship using *area homologs,* demonstrate ways to deal with geographical paralogy and MASTs, and resolve areagrams and extract subtrees using paralogy-free subtree analysis and the transparent method.

We also discuss ways to combine subtrees to find the minimal tree or *general areagram* using three-item and compatibility analysis.

Finally, we explore ways to interpret geographical congruence using the Mexican Transition Zone as an example. The importance of hierarchy analysis and time in general areagrams is also emphasized.

measures biotic or geographical proximity based solely on the shared taxic composition of areas is a similarity, or *phenetic,* method. Without biotic—that is, taxic—homology, there is no concept of evolutionary relationship. In Chapter 6, we reviewed a range of techniques that use biotic and geological similarity to establish area classifications. Those methods are considered preliminary in establishing a biogeographical classification because their results need to be tested using biotic homology (geographical congruence) in a systematic biogeographical analysis. Here, we describe a method for proposing biogeographical classifications based on the distribution and relationships of biotic areas, incorporating geological and geographical data.

Describing and Diagnosing the Study Area

All biogeographic studies require careful identification of the study area(s) and the areagrams that are appropriate for resolving relationships among included endemic areas. The overlap of taxic distributions delimits the study area within a specified geographical range: the areas of endemism (see Chapter 3). Donn Eric Rosen pioneered this method using Central American freshwater fishes of the family Poeciliidae (Rosen, 1978, 1979). Rosen's method is straightforward: choose the taxa for which you have robust phylogenetic hypotheses, and then use their distributions to delimit the areas of endemism to be studied.

Maps are one of the best ways to express and communicate what we mean by areas of endemism. Rosen (1978) used maps to discover

Box 7.1 Donn Eric Rosen (1929–1986)

Figure 7.1. Donn Rosen in his office and lab at the American Museum of Natural History. [Photograph courtesy of the late Carmela B. Rosen.]

Donn Eric Rosen was an ichthyologist at the American Museum of Natural History for most of his career, which ended abruptly with his death at age 57 (see Nelson et al., 1987). Donn was a native New Yorker and the younger brother of Charles Rosen, a renowned classical pianist, musicologist, and literary critic.

Donn Rosen's influential studies of the systematics and biogeography of livebearing fishes of the genera *Xiphophorus* and *Heterandria* (family Poeciliidae) quickly became benchmarks of the application of cladistic methods in biogeography. AMNH colleague, ichthyologist, and biogeographer Gareth Nelson; the late British paleontologist Colin Patterson; and the late British ichthyologist P. Humphry Greenwood were among his closest collaborators.

Rosen was a gifted teacher and, through his appointment as an adjunct professor at the City University of New York, served as major professor to a series of doctoral students, notably for biogeography including E. O. Wiley, Richard Vari, Lance Grande, and Lynne Parenti. Despite their philosophical differences, especially over the significance of the biological species concept, Rosen and Ernst Mayr, a curator at the AMNH before becoming a professor at Harvard University, maintained a cordial relationship. Rosen married Carmela Berritto, a former classmate at New York University and Mayr's research assistant at the AMNH, who co-authored a study of geographic variation in Bahamian snails (Mayr and Rosen, 1956; see Chapter 2).

Figure 7.2. Marron, *Cherax tenuimanus*, from southwest Western Australia. [Photograph copyright of Georgina Steytler.]

Figure 7.3. Red-winged fairy wren, *Malurus elegans*, from southwest Western Australia. [Photograph copyright of Georgina Steytler.]

Figure 7.4. Karri trees, *Eucalyptus diversicolor*, from Northcliffe, southwest Western Australia. [Photograph copyright of Tony Windberg.]

areas of endemism by drawing overlapping distributions of taxa. The same maps can also be used to propose the limits of these areas. Not surprisingly, as we noted in Chapter 3, endemic areas overlap in space and in time. As an example, endemic distributions of a freshwater crustacean, the marron *(Cherax tenuimanus)* (Figure 7.2); a passerine bird, the red-winged fairy wren *(Malurus elegans)* (Figure 7.3); and the karri tree *(Eucalyptus diversicolor)* (Figure 7.4), one of the iconic Australian eucalypts, overlap in the southwestern region of Western Australia and form a biota (Figure 7.5). Once recognized, the biotic area is diagnosed by its taxa and its inorganic features.

An area of overlap reflects a common biotic distribution. Biotic distributions alone are insufficient to diagnose the area. To diagnose the area properly, we need to hypothesize what inorganic disjunctions

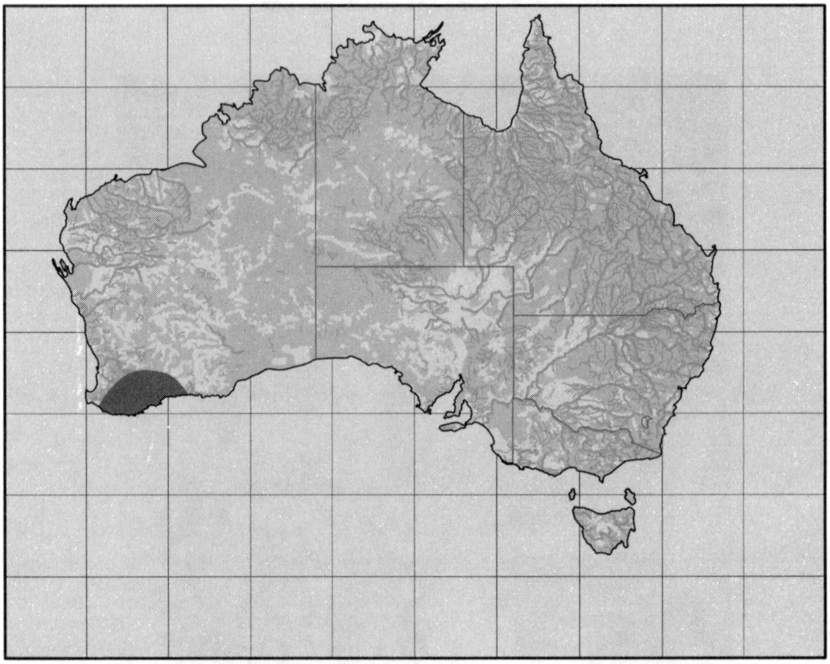

Figure 7.5a. Distributional range (green) of marron, *Cherax tenuimanus*, from southwest Western Australia.

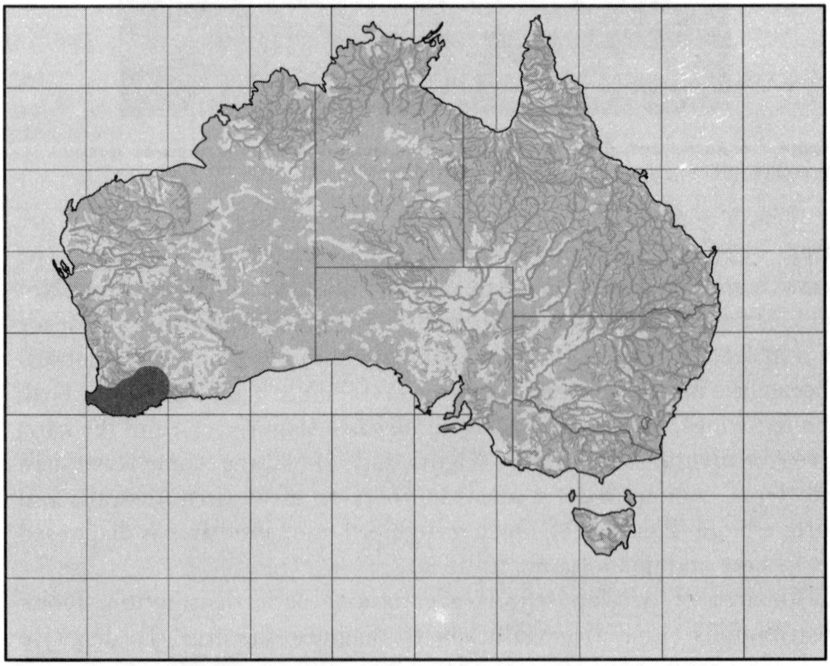

Figure 7.5b. Distributional range (green) of the red-winged fairy wren, *Malurus elegans*, from southwest Western Australia.

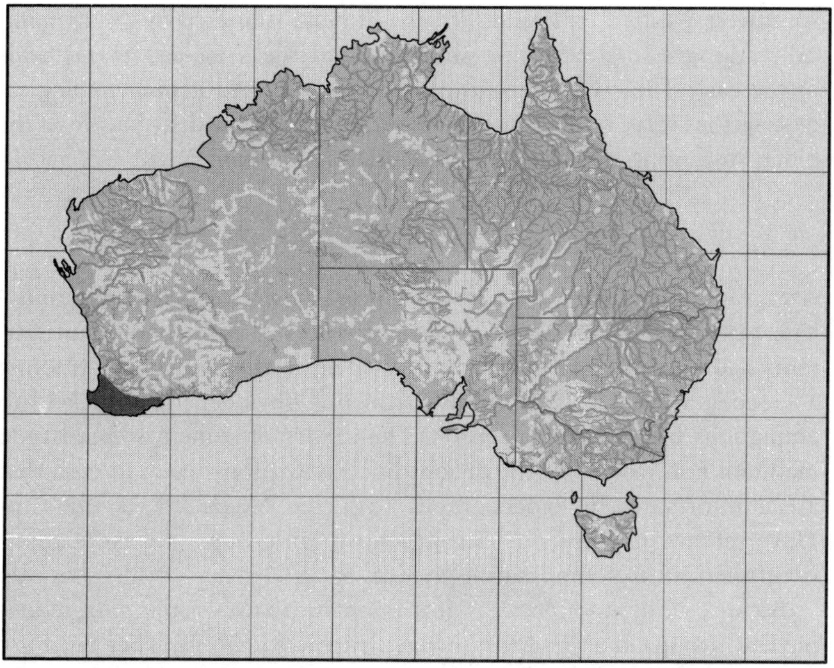

Figure 7.5c. Distributional range (green) of the karri tree, *Eucalyptus diversicolor*, from southwest Western Australia.

correspond with the common distributional pattern: Are the distributions limited to a particular climatic range? Do seaways or any other geological limits form any border of the distributions? Are the distributions hypothesized to have been formed by a geological event, such as a flood, earthquake, tsunami, or retreat of a glacier?

The marron lives in cool freshwater streams that are high in oxygen and low in suspended sediment. The red-winged fairy wren requires a cool-temperature forest with low cover, and the karri tree needs loamy soil, consisting of clay, sand, and organic debris, with high rainfall. The southwestern region of Western Australia is in a cool-temperate climate characterized by loamy soil with a high groundwater level. The soil chemistry and underlying geology change gradually toward the east, where the soil is harder, less moist, and more acidic. Toward the west, the soil is sandy with a high groundwater level that forms bogs and inlets. Fossils representing lineages of extant bivalves and gastropods that lived further inland suggest that there have been several marine transgressions and regressions that were possibly responsible for the loamy soil. The southwest Western Australian terrestrial biota is limited by the dynamically transgressing/regressing shoreline and the arid, less loamy soils of the east. These basic geographical, climatic, and geomorphological descriptions serve to diagnose the natural limits of the distribution of the biota, suggesting its former and potential distribution range.

Choosing Areagrams

Areagrams for comparative analysis *must* be derived from monophyletic taxa. Non-monophyletic "groups" do not form congruent patterns and are therefore uninformative or misleading when constructing a biogeographical classification. Erroneous phylogenetic data lead to ambiguous biogeographic results. The efforts of some systematists to maintain non-monophyletic groups in classifications result in taxa that are uninformative in biogeography (e.g., the "Algae," "Reptilia," or "Invertebrata"; Chapter 1). Retention of non-groups in classifications is antithetical to a comparative science.

Because we look for patterns, it is axiomatic to use overlapping, monophyletic groups in a systematic biogeographical analysis. They need not represent taxa of the same age, as might be estimated through fossils, molecular clocks, or stratigraphy. Estimates of age of all or portions of a biota can be used to interpret biogeographic patterns. We explain this below in our discussion of hierarchy analysis.

Geographical Overlap

With overlapping areas of endemism, even if the degree of overlap is relatively small, we are able to compare the resultant areagrams. At least two areas should overlap (Figure 7.6). Two areas overlapping ensures that a relationship can be established (Chapter 3). The smallest unit of relationship is two areas that are more closely related to each other than either is to a third area: a three-area relationship or area homolog.

Temporal Overlap

Areas and the biota that inhabit them change over time. The Australian continent did not have the same fauna and flora 50 mya as it does today. Therefore, the biotic areas that we define for the purposes of a comparative biogeography of Australia of 50 mya are not identical to the areas today, *but they could express the same area homolog or area relationships.*

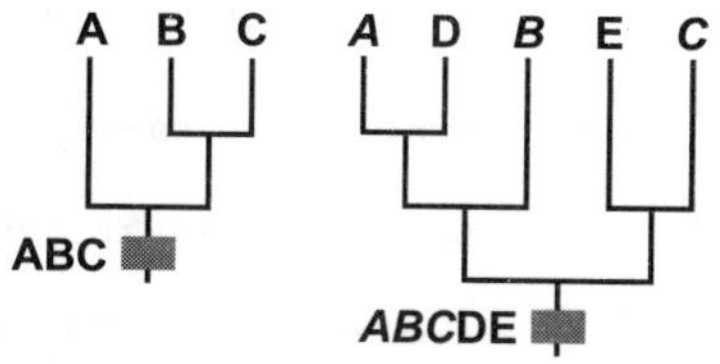

Figure 7.6. Area overlap between two clades. Areas A, B, and C overlap.

Temporal overlap is critical for non-paleontological as well as paleontological biotic areas. The geological areas of Southeast Asia, for example (see Hall, 1996, 2002), are complex and have evolved rapidly, meaning that older taxa may express different area relationships than younger taxa. Temporal overlap can cause conflicting biogeographic patterns recognized at different hierarchical levels. We address ways to recognize and interpret this conflict below. It requires recognizing the distinction between areagrams and TACs, which we review first.

THE STRUCTURE OF AREAGRAMS AND TAXON-AREA CLADOGRAMS

Areagrams (area cladograms) and *taxon-area cladograms* (TACs) are two types of branching diagrams used to represent different types of biogeographical information. We introduced some of the concepts of a comparative biogeography in Chapter 4; all of the area relationships were interpreted as areagrams. In Chapter 5, we introduced the differences between areagrams and TACs. In Chapter 6, we discussed many cladogram-based biogeographic methods that use TACs. Here, we expand on the significant differences between areagrams and TACs and explore how these differences affect biogeographic analyses.

Areagrams and TACs are comparable to cladograms and phylogenetic trees, respectively, in systematics. An areagram is a classification of endemic areas based on area homologs, whereas a TAC is a phylogenetic tree that includes geographical information on endemic areas. Areagrams are summaries of relationships among areas. Unlike TACs, areagrams do not represent or contain information on any kind of optimization, be it of characters, ecology, distributional mechanisms, centers of origin, or anything else.

These differences between areagrams and TACs have been overlooked in most comparisons and contrasts among biogeographic

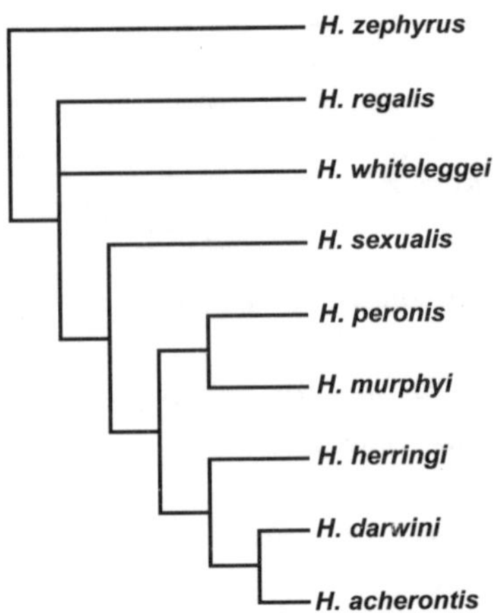

Figure 7.7. A cladogram of nine species of marine water-striders, genus *Halobates* (after Andersen, 1998: Figure 2a).

methods. Yet their differences are fundamental. The apparent misunderstanding of the differences between areagrams and TACs stems from different goals of biogeographic analyses. These have led to different methods for deriving branching diagrams from taxon cladograms and phylogenetic trees.

CONVERTING CLADOGRAMS AND TREES INTO AREAGRAMS AND TACS

To demonstrate how to convert phylogenetic hypotheses into areagrams and TACs, we use the phylogenetic hypothesis of Indo-Pacific species of marine water-striders (Heteroptera, Gerromorpha) of *Halobates zephyrus* and the *H. regalis* group, as proposed by Andersen (1998). A well-resolved hypothesis of relationships among nine water-strider species is represented in a cladogram (Figure 7.7). This hypothesis is congruent with that of Andersen (1991; Chapter 4) with the addition of *H. murphyi* from New Guinea, as sister species of *H. peronis*.

The nine water-strider species live in four broad areas of endemism, as defined by Andersen (1998:344): Australia, Malaya, Papuasia, and the Philippines. To convert the phylogenetic tree of Figure 7.7 into an areagram, the name of the taxon is replaced by the name of the endemic area in which it lives (Figure 7.8; see also Andersen, 1998: Figure 2). *No other inference* is transferred or translated onto the topology of the areagram.

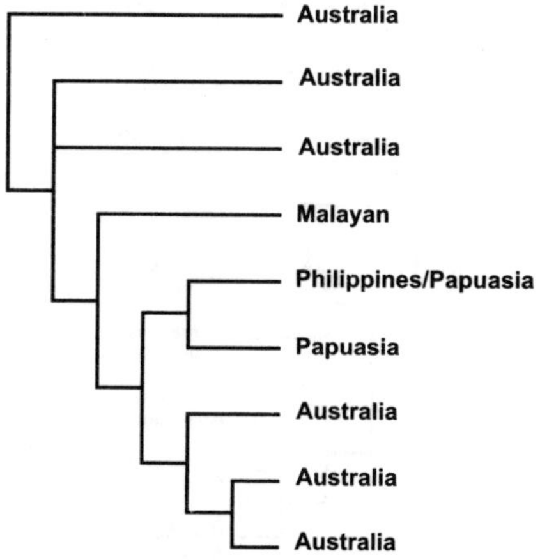

Figure 7.8. The areagram of water-striders, derived by replacing the name of the taxon in Figure 7.7 with the name of the endemic area in which it lives, following Andersen (1998).

To convert the same phylogenetic tree into a TAC, *all the information inferred* to be included in the phylogenetic tree must be transferred to the topology of the TAC (Figure 7.9). This is a critical distinction that we have introduced in previous chapters and that we explain further here.

The differences between cladograms and phylogenetic trees were debated extensively in the phylogenetic systematic literature during the height of the cladistics revolution (e.g., Cracraft, 1979; Platnick, 1979; Wiley, 1979). It was well appreciated that cladograms represent the distribution of characters among taxa and that phylogenetic trees represent that same distribution of characters with the added assumption of transformation. This distinction has been absent from many of the recent discussions of phylogenetic analysis.

The differences between the areagram and the TAC in biogeographic analysis are analogous to those between the cladogram and the phylogenetic tree. As between cladograms and phylogenetic trees, the differences are of interpretation, not of topology. In the areagram (Figure 7.8), the geographically paralogous terminal branches labeled "Australia," for example, are *the same branch*. There is no difference between them, as they represent the area "Australia" and *nothing else*. The taxa that represent this geographic paralogy are not present in the terminal branches of an areagram. Therefore, only one branch termed "Australia" is considered unique; the others are duplicates that present redundant information.

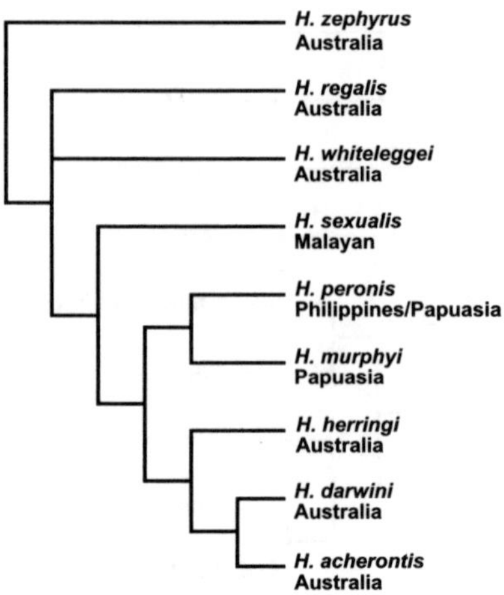

Figure 7.9. The TAC derived from the cladogram of Figure 7.7 by adding the endemic area in which each taxon lives.

In contrast, the TAC (Figure 7.9) contains *all* the information about taxa and the transformations among them that is inferred to be present in the original phylogenetic tree. Each and every terminal branch is unique, whether or not it is geographically paralogous or repeated. Each of the terminal branches labeled "Australia" on the TAC is different because it represents a different taxon—here, species of water-striders. The difference between areagrams and TACs is most profound at the level of nodes and components.

Ancestral Nodes and Components

The concept of nodes in biogeography follows from that in phylogenetic systematics.[1] In biogeography, nodes on a cladogram are treated as being either one of two concepts: *ancestral nodes* or *components*.

Ancestral nodes are the nodes of a cladogram or phylogenetic tree that are interpreted to represent an ancestral taxon, an ancestral area, or an evolutionary, distributional, geographical, or physiological mechanism (e.g., sympatry, climate, or physiology favorable for wind dispersal or rafting, and so on). Ancestral nodes can also be interpreted as representing particular historical events, such as island-hopping from an older to a more recently formed volcanic island (e.g., Wagner and Funk, 1995) or responding to Alpine glaciations (e.g., Schönswetter et al., 2002).

Ancestral nodes are crucial for many different types of evolutionary biogeographical analyses, such as Brooks parsimony analysis (BPA),

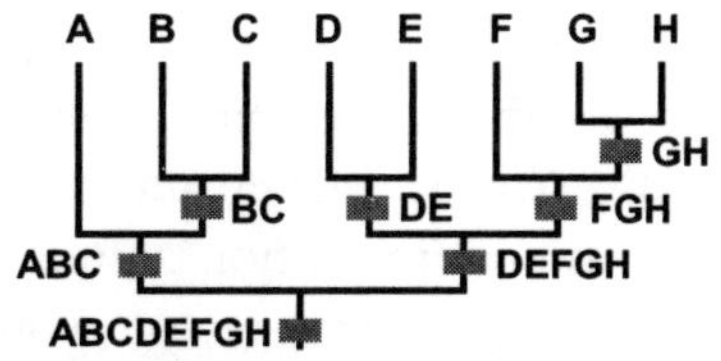

Figure 7.10. Components represent topology.

modified BPA, PACT, ancestral area analysis, DIVA, and comparative phylogeography (Chapter 6). In these analyses, the ancestral node is interpreted as evidence of an evolutionary, distributional, or geographical hypothesis, such as a center of origin, which generates an explanation, such as dispersal from that center of origin. In BPA, ancestral nodes act as ancestral taxa. In AAA and DIVA, ancestral nodes represent and act as ancestral areas and hypothesized former distributional mechanisms, respectively. Ancestral nodes are found exclusively in TACs. Ancestral nodes are not present in areagrams. In areagrams, nodes are interpreted as components.

Components are junctions on branching diagrams that represent the content of the terminal branches. They contain no explicit information about evolutionary, distributional, geographical, or physiological mechanisms or events that formed the node or led to its subsequent diversification. Components simply indicate relationships among areas (Figure 7.10). Explanation of these relationships by speculating on a biogeographic mechanism, however compelling, is secondary.

Again, whereas TACs and areagrams are structurally identical, they are fundamentally different at the level of nodes and components. Without the interpretation of nodes as ancestral, TACs *are* areagrams. Components are defined by the terminal relationships within an areagram, whereas ancestral nodes characterize TACs. The interpretation of each ancestral node forms the basis of the hypothetical scenarios and mechanisms that are proposed to explain TACs. Areagrams form components based on the content of each terminal branch. The information that components convey rests on their interpretation.

The difference between areagrams and TACs is usually expressed during the interpretation of distributional history. Someone interpreting the cladogram of Figure 7.9 as a TAC might conclude that water-striders originated in Australia and, after periods of differentiation, colonized Malaya, then moved to Papuasia and the Philippines, and finally dispersed back to Australia. Someone else interpreting Figure 7.9 as an areagram would say that Australia is related as in the subtree ((Papuasia, Philippines)Australia)Malaya and that the duplication of areas on the areagram adds no information to, nor does it detract from, that set of area relationships. The first person would conclude that he has interpreted the distributional history of the water-striders. The second person would say that she has discovered a statement of area relationships to be tested with data from other taxa to determine whether or not the water-striders are part of a general distribution pattern or biogeographical classification. If the water-striders are part of a general distribution pattern, then a general explanation can be sought for the pattern. If the water-striders have a unique distribution pattern, then a unique explanation may account for their distributional history. The first person is operating within an explicit center of origin/dispersalist paradigm, whereas the second person has not specified a preference for dispersal or vicariance as an explanatory mechanism.

Choosing between Areagrams and TACs: Classification or Explanation?

The cladistic or phylogenetic focus of systematics gives us a clear choice: do we use areagrams to discover area homology and classifications, or do we use TACs to form hypotheses based on select distributional paradigms?

Our choice depends upon our assumptions. Our assumptions, like the methods we use to hypothesize phylogenetic trees that we convert into TACs or cladograms that we convert into areagrams, influence the way we do biogeography—the way we think about biological distributions. The way we do biogeography reflects the way we approach comparative biology. Our assumptions likely do not change as we transition from systematist to biogeographer. When a single node on a cladogram can be interpreted in different ways, depending on which method one chooses, biogeography is a highly subjective field (see Figure 7.11). More important, most biogeographers have not thought about the differences between areagrams and TACs and do not realize how profoundly their biogeographic analyses are directed by these unspoken assumptions.

Before we decide how to convert our cladograms or phylogenetic trees into branching diagrams for the purpose of biogeographic analysis, we consider what that conversion actually means. What sort of data are we

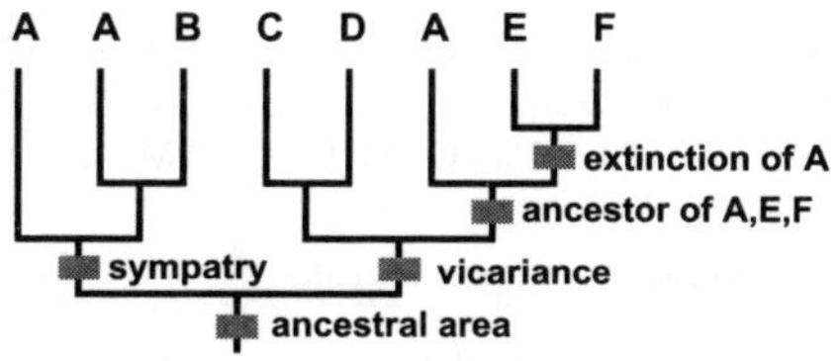

Figure 7.11. Nodes of TACs are given meaning according to the particular model applied.

transferring when we convert a branching tree based on taxic characteristics into another that has ancestral nodes or components?

The Real World: What Do We Convert?

The phylogenetic pattern we uncover is biased by our choice of characteristics and their character states or alignments. The trees and cladograms we discover change depending on how we choose to treat the characters and character states (as transformations, as relationships, or as similarities). Different methods mean different trees. Even the way we select among our trees results in different consensus trees. At any time during this process, we may introduce error. Any cladogram, phylogenetic tree, or gene tree contains a certain degree of error. We may include a homoplastic character or a non-monophyletic taxon. We unwittingly overlook an informative morphological character or align our sequences incorrectly. There are many errors and biases inherent in cladograms and trees that we convert into areagrams and TACs, highlighting the seemingly imprecise nature of biogeography (Table 7.1). What do we hope to glean from such possibly incongruent data?

Biogeographers seek congruent biogeographical hypotheses. This is not realized in every biogeographic analysis. The world changes biologically as much as it does geographically and geologically. We likely will never know the precise history or the past events or mechanisms that caused our distribution patterns. We are limited to proposing processes based on congruent patterns. This seems basic, but perhaps too restrictive, for some biogeographers who wish to interpret the TAC of a single taxon, as would the first person in our example above with the water-striders. As systematists, we should accept the unresolved nodes that riddle our trees and cladograms and acknowledge that both morphological characters and molecular sequences are difficult to extract and even more difficult, or even impossible, to uncover the further we move back in geological time or in evolutionary history.

TABLE 7.1 SOME CAUSES OF AMBIGUITY IN CONVERTING CLADOGRAMS, PHYLOGENETIC TREES, AND GENE TREES INTO AREAGRAMS AND TACS

Polytomies	Geographical Paralogy	MASTs
Unresolved cladograms/trees	Poor area definition (area too large)	Poor area definition (overlapping areas)
Too few characters in systematic analysis	Poor taxon sampling (e.g., distribution based on single specimen)	Poor taxon sampling (e.g., distribution based on non-monophyletic group/s)
Conflicting taxic characters	Temporal distributions not included	Temporal distributions not included

Norman Platnick (1991:ii) bluntly summarized reality for both systematists and biogeographers:

> By the time we have done the fieldwork to sample populations, collected the data to discriminate species, cladistically analysed their interrelationships, and mapped their distributions, the group of organisms in question gains a heavy burden of anticipation of biogeographically decisive resolution.
>
> In systematics, there is no *a priori* reason to expect that a given character system (be it morphological, physiological, behavioural, or molecular) will provide data crucial to the resolution of cladistic relationships at some particular hierarchical level. Is there an *a priori* reason to expect that the distributions and interrelationships of a given group of taxa will provide data crucial to the resolution of area relationships in some particular part of the world?

Given this, it sometimes seems prudent to model historical events or explanatory mechanisms, but such modeling does not guarantee any more rigor or explanatory power.

What we convert from cladograms and phylogenetic and gene trees thus includes much incongruence and ambiguity. In every systematic biogeographical analysis, there are areagrams that are possibly incongruent or that contain poorly defined endemic areas. Congruence is rare. We discover congruence by recognizing and removing redundant data and resolving incongruent relationships.

SOLVING SINGLE AREAGRAMS

Ambiguity in areagrams appears as geographical paralogy, MASTs, or polytomies. In Chapter 4, we showed how these ambiguities convey little information about area relationships, or, in the case of MASTs,

Box 7.2 Geographical Congruence and General Areagrams

When two or more sets of overlapping cladistic relationships are congruent, they form a pattern, which signifies geographical congruence. Identifying geographical congruence is the goal of systematic biogeography. Geographical congruence is represented by a common pattern for all areagrams in an analysis. Geographical congruence may be used to interpret the biotic history of a region or area of endemism.

Geographical congruence is found in general areagrams. General areagrams contain components that represent biotic divergence. The mechanisms driving biotic divergence are unspecified. As a classification of biota, a general areagram does not endorse any one scenario (e.g., vicariance versus dispersal) over another.

how they may contain potential information on relationships. Any areagrams that contain such ambiguities must be resolved. In systematic biogeography, we compare congruence, or patterns, not incongruence, or non-patterns. Here, we provide a way to address ambiguity.

Basic Methodology

A single areagram is *not* a pattern. Cladograms and phylogenetic and gene trees are expressions of distributions of characters that form a classification: a hierarchy of homologous character relationships. In systematic biogeography, a single areagram is analogous to a character or character-state tree and therefore is merely a single statement of relationship (one or a set of area homologs) that may form a pattern with other, congruent areagrams (Figure 7.12).

An areagram does not represent a biogeographical pattern if it contains only a single proposal of area homology that has yet to be corroborated with biogeographic data from other clades.

The endemic areas that we test for area homology and the larger areas, regions, and realms that we test for area monophyly are expressed as a relationship. Areagrams contain hypotheses of potentially congruent area relationships and may be congruent with other areagrams.

Furthermore, areagrams are a collection or set of different area homologs. To express these area homologs unambiguously, we need to resolve MASTs and remove polytomies and geographic paralogy—in effect, to form "reduced" areagrams, a concept introduced by Rosen

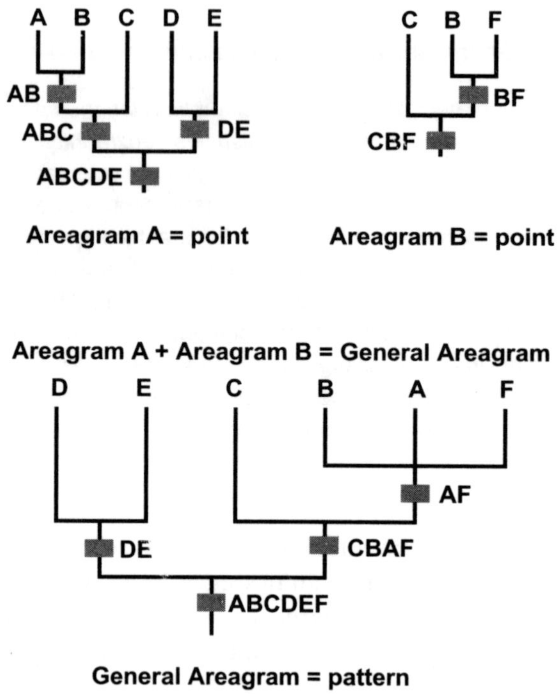

Figure 7.12. The difference between points and patterns. Points are single occurrences of relationship. Patterns are congruent relationships shared by areagrams.

(1978, 1979, *sensu* Reduced Area Cladograms). Reduced areagrams do not contain any redundant or ambiguous information about area relationships.

Paralogy-Free Subtree Analysis: Treating Geographic Paralogy and MASTs

The paralogy-free subtree method is appropriate for all areagrams that contain geographic paralogy and MASTs (see Chapter 4). The ingenuity of the subtree method is that it deals with all data purely by way of relationship. In dealing strictly with relationship among areas, the method treats ambiguity as uninformative after recovering any potentially informative area relationships. The method is free of *a priori* evolutionary, biogeographical, or geographical models.

Paralogy-free subtree analysis has been called an *intuitive* method (see Ebach et al., 2005) because it involves recognizing area homologs among an array of informative and uninformative nodes, rather than

Box 7.3 MASTs and Information

Areagram with non-paralogous MAST

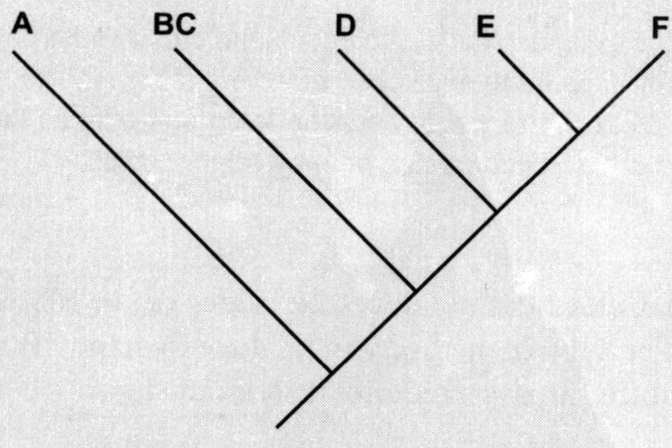

specifies two, fully resolved areagrams:

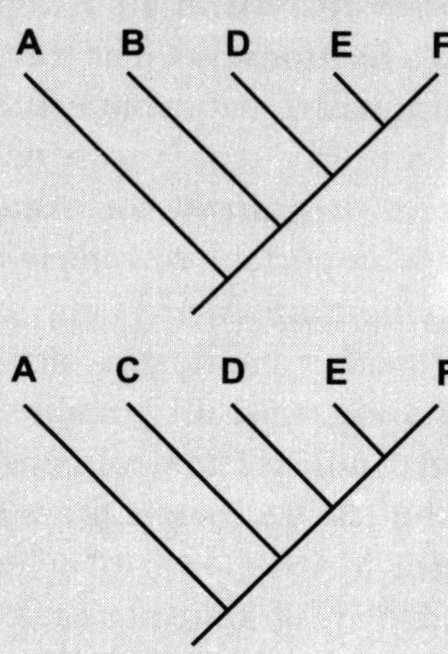

Figure 7.13. Areagram with non-paralogous MAST (BC), above, includes two fully resolved areagrams, below.

MASTs are not always geographically paralogous and may contain information. A MAST without geographically paralogous areas can contain different area relationships:

A(BC(D(EF))) = A(B(D(EF))) + A(C(D(EF)))

MAST BC is not paralogous and provides information on the relationships of areas B and C. *(continued)*

In some cases, MASTs contain no information on area relationships:

A(B(C(DE(DE)))) = A(B(C(DE)))

MAST DE is geographically paralogous and contains no potential information. It *appears* to have been removed in the resolved areagram A(B(C(DE))). DE is not repeated because to do so would be redundant (i.e., it would not add information on area relationships).

proposing that all nodes are ancestral nodes (as in BPA or DIVA; see Chapter 6). The subtree method can be done by hand. Translated as an algorithm, subtree analysis requires a series of steps.

Implementing Subtree Analysis: The Transparent Method

The first stage of subtree analysis of an areagram with MASTs is to identify all possible area relationships. The transparent method uncovers any potentially informative data contained in MASTs by treating each occurrence of an area in a separate areagram. An areagram with one MAST that contains two areas, for example, can be expressed as two areagrams. It is important to remember that areagrams are *expressions of area relationship*. MASTs may contain several possible relationships that are neither paralogous nor conflicting. By depicting these as separate areagrams, all possible area relationships can be found, and potential information recognized and extracted. The areagram of satinfin shiners, freshwater fishes of the genus *Cyprinella* (Figure 7.14), has three MASTs, each of which has two areas. The total number of possible areagrams, not each of which is necessarily informative, is 2^3, or 8.

MASTs *may also contain geographically paralogous areas*. These are removed in the next stage.

Implementing Subtree Analysis: Paralogy-Free Subtrees

The essence of paralogy-free subtree analysis is to uncover information content—the area relationships within an areagram—by relating unique sets of areas called *subtrees*. Repetition within a subtree is geographic paralogy. An areagram may itself be a subtree or may contain two or

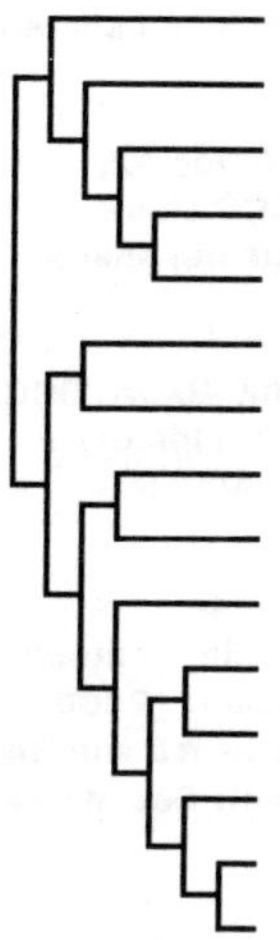

N	N	N	N	N	N	N	N	N
N	N	N	N	N	N	N	N	N
3, N	3	N	N	N	3	3	N	3
N	N	N	N	N	N	N	N	N
N	N	N	N	N	N	N	N	N
3	3	3	3	3	3	3	3	3
5	5	5	5	5	5	5	5	5
3	3	3	3	3	3	3	3	3
3	3	3	3	3	3	3	3	3
3	3	3	3	3	3	3	3	3
3	3	3	3	3	3	3	3	3
3	3	3	3	3	3	3	3	3
1,3	1	1	1	3	3	3	3	1
3	3	3	3	3	3	3	3	3
3,5	5	3	5	5	3	5	3	3

Figure 7.14. The areagram of satinfin shiners, freshwater fishes of the genus *Cyprinella* (redrawn from Marshall and Liebherr [2000: Figure 2], in part, following analysis of Mayden [1989]) and its eight possible MAST-free areagrams (box), not each of which is equally informative. See Marshall and Liebherr (2000) for further identification of areas.

Box 7.4 Rules for Finding Subtrees

1. Subtrees are unique sets of area relationships.
2. Two subtrees in the same areagram may contain the same areas but can have different relationships among those areas.
3. A subtree "stops" (i.e., is complete) when another set of unique relationships begins at an internal node.

more subtrees. A MAST-free areagram may contain geographically paralogous areas. Geographically paralogous areas are essentially copies, counter to the argument that geographic paralogy is meaningful and its apparent removal means discarding useful data.[2] Geographic paralogy is uninformative, and its removal is expressed *figuratively*. The geographically paralogous areas are visibly removed, but the area relationships, the information content of the areagram, cannot be affected by their removal. As above, if we view the areagram as a relationship, then geographically paralogous areas have no individual identity. They are only repeated caricatures, letters, names, figures, or drawings on a branching diagram. The areagram (Figure 7.8) specifies the subtree

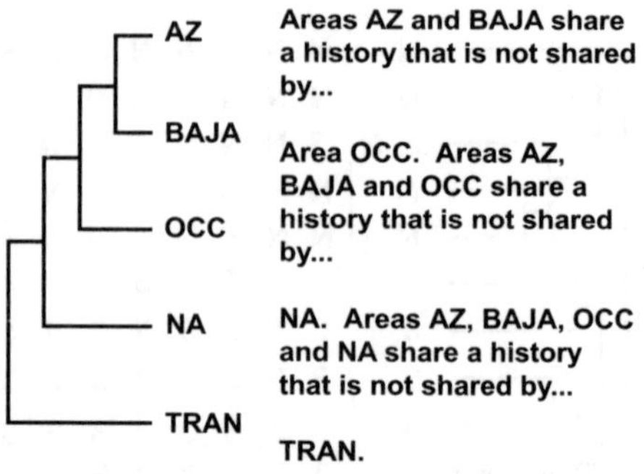

Figure 7.15. Description of relationships of areas within an areagram of beetles of the genus *Typhlusechus* (modified from Marshall and Liebherr [2000: Figure 2], in part, following analysis of Aalbu and Andrews [1985]). Areas AZ and BAJA are at the "top" of the areagram (i.e., they are the most deeply nested). See Marshall and Liebherr (2000) for further identification of areas.

((Philippines, Papuasia)Australia)Malaya. The repeated occurrence of "Australia" is geographic paralogy, adds nothing to the information content of the areagram, and is removed.

Subtree analysis has a purely comparative goal: relationship. To relate one thing to another, we compare. In areagrams, comparison starts at the top-most or cladistically nested components, and the relationships may be described in the same way that we describe relationships among taxa (Figure 7.15).

If the areagram contains one or more subtrees, indicated by a bifurcating node or component, then each subtree is treated as a separate areagram that may be informative, if it contains more than two areas, or uninformative if it contains only two (Figure 7.16).

Subtree Analysis and Polytomies

Subtree analysis cannot resolve polytomies. We may include polytomies in the analysis, but once areagrams are combined, the polytomies will logically reappear. If we remove polytomies prior to a subtree analysis, the result is unaffected. Non-paralogous areas that are included in any polytomies will automatically be removed and will not appear in any recombined areagram or general areagram.

Computer Programs to Implement Subtree Analysis

Nelson05 is the computer program that fully implements subtree analysis (Ducasse et al., 2007).[3] TAS (Nelson and Ladiges, 1991a) and TASS (Nelson and Ladiges, 1994, 1995), earlier programs to

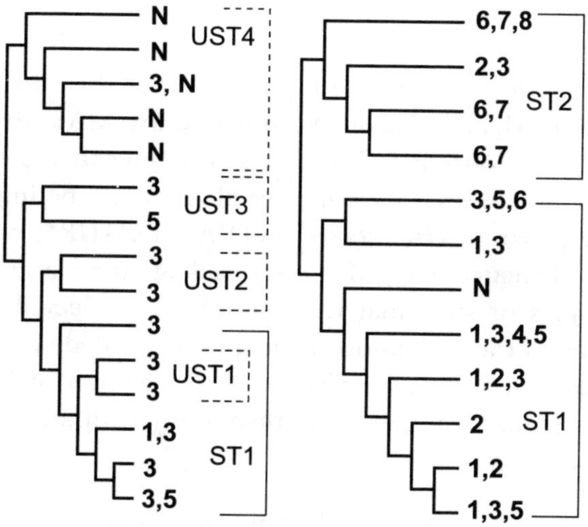

Figure 7.16. Subtrees identified in areagrams of satinfin shiners, freshwater fishes of the genus *Cyprinella* (left), and horned lizards, *Phrynosoma* (right; redrawn from Marshall and Liebherr [2000: Figure 2], in part, following analysis of Montanucci [1987]). Uninformative subtrees (UST) are indicated by dashed brackets, informative subtrees (ST) by solid brackets. See Marshall and Liebherr (2000) for further identification of areas.

Box 7.5 Subtree Analysis Step-by-Step

Subtree analysis can be done by hand—first by relating areas within an areagram and then by finding any subtrees.

Step 1: Start from the top-most (most highly nested) area of any internal component or node.

Step 2: Relate the top-most area with the succeeding areas.

Step 3: If an area is geographically paralogous, it will not relate to itself and therefore should be discounted.

Step 4: If a MAST is encountered, relate the first non-paralogous area.

Step 5: Once you reach an internal node, relate all non-paralogous area relationships, unless the internal node signifies the end of the subtree.

Step 6: If the subtree contains a MAST with additional non-paralogous areas, start again at step 1 and include these to form all possible area relationships.

You should discover one set of relationships, and more if MASTs or subtrees are present.

You may add the subtrees or areagrams by hand or using a consensus or minimal tree method (see Box 7.6).

implement subtree analysis, are limited, in that they do not resolve non-paralogous MASTs. For areagrams that are MAST-free, TASS is an alternative to *Nelson05*. But TASS is unable to recombine subtrees; instead, it creates a data matrix file that can be imported into a parsimony program (*Hennig86, NONA,* or *PAUP**) or any other cladistics or phenetics (e.g., *MrBayes*) package for manipulation. Of course, analyses of such matrices should be conducted with a full understanding of the assumptions being implemented, as noted above and in Chapter 6. Below, we discuss an application of *Nelson05* to the Mexican Transition Zone, and then apply it to areagrams of the Pacific in Chapter 9.

MAST-Riddled Areagrams

Consider two groups of species living in areas A, B, and C. Each areagram contains multiple MASTs: ABC(ABC, ABC) and AC(C, ABC). That is, the first group has three species that each live in all three areas, and the second group has three species, two of which have more restricted distributions. What area relationships do these areagrams contain?

1. The areagram ABC(ABC, ABC) contains three informative three-area statements or area homologs, A(CB), B(AC), and C(AB), which, when added together, form the trichotomy ABC.
2. The areagram AC(C, ABC) contains one informative three-area statement or area homolog: A(CB).

This example demonstrates that MAST-riddled terminal branches may either (1) generate a number of possible relationships that conflict (because MASTs can only be resolved by way of relationship, no single informative area homology can be discovered if there is a series of possible, conflicting relationships) or (2) resolve as a single, unique area relationship (that is, one area homolog is *discovered*).

MAST-riddled areagrams are likely to generate many conflicting relationships that add no informative data to the overall analysis. To reduce the number of conflicting areagrams in an analysis, MAST-riddled areagrams that produce many conflicting relationships should be identified and removed from further consideration in that analysis. They should not be discarded completely because they may provide data on area relationships in other studies and support the interpretation of general areagrams. Also, areas could be redefined so that they are biogeographically meaningful.

THE GENERAL AREAGRAM: COMBINING SUBTREES AND AREAGRAMS

A general areagram is a summary of all the information contained in the areagrams and subtrees. There are various ways to combine MAST-free and geographic paralogy-free areagrams and subtrees. In many biogeographical methods, finding a consensus areagram is based on a model of parsimony, maximum likelihood, Bayesian analysis, or compatibility. The input data, be it binary or in the form of trees, can be manipulated in different ways. We do not rely on a particular consensus method to establish a general areagram, as each of the many consensus methods, such as strict or majority-rule consensus, may produce a different consensus areagram. Instead, systematic biogeography seeks a way in which to express the congruent relationships contained in each areagram to find the general areagram—the branching diagram that best expresses *geographical congruence*.

Three-Item Analysis

To discover the general areagram, we list all the area homologs that occur in the input areagrams and subtrees. The smallest statement of relationship is a three-item statement; three-item analysis (Nelson and Ladiges, 1991b, 1991c; Nelson and Platnick, 1991) is the method we use to uncover the general areagram. Because we aim to compare area homologs equally, we use three-item analysis, which is a rigorous empirical method.[4]

Three-Item Analysis and Area Homologs (Characters)

Three-item analysis does not interpret distributional data as character states in a transformation series. Instead, all character states are related based on an *input hierarchical order* as discovered by the biogeographer. The input hierarchy is tested using three-item analysis. From a transformational point of view, this *appears* to be "ordering character states," although no transformation, from one character or character state to another, is used or assumed.

From the alternative point of view, a homolog is a *statement of relationship*—with relationship being *hierarchical* and *not transformational*. The hypothesis of relationship is being tested. Presence of a homolog (character) in two organisms relates these two organisms when it is *compared to a third* in which the homolog is absent. Our

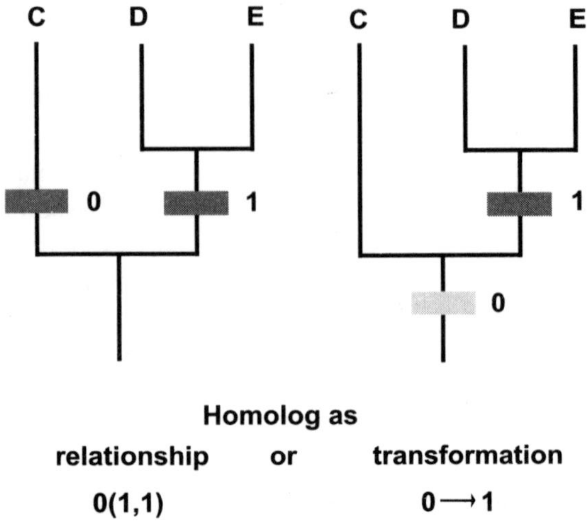

Figure 7.17. A homolog, C(DE), as a relationship versus a hypothetical transformation.

hypothesis, therefore, is that these organisms are related by that particular homolog when compared to others (Figure 7.17).

By inputting hierarchies, three-item analysis is a directed analysis. In this sense, three-item analysis is a universal evolutionary method, as it accommodates all possible evolutionary models. Because three-item analysis does not discriminate among evolutionary models, it will not favor any mechanism over another. The search for homologies makes three-item analysis evolutionary: it accommodates change over time.

Once hierarchies (Figure 7.18) are entered using the three-item software *Nelson05,* the program finds the general areagram that includes all relationships from the input subtrees and areagrams that are the most resolved. Before the program does this, it assesses the nature of the data—finding how many relationships there are and if these are equally represented.

Three-Item Fractional Weighting

Three-item weighting results in a fractional value that is the number of times an independent, informative statement occurs compared to all possible informative statements. It should not be confused with the practice of weighting the relative importance of characters in statistical and phenetic systematic methods. In the areagram (ABC)DE, there are four independent, informative three-item statements: D(AB), D(AC), E(AB), and E(AC). When added together, they recover the areagram (ABC)DE.

HOMOLOGS

Character 1: *Forelimbs*	Character 2: *Skin covering*
States: *fin, wing, forearm*	States: *scales, feather, hair*

HIERARCHY

minnow (wren, human)

HOMOLOGIES

Character 1: fin (wing, forearm)

Character 2: scales (feather, hair)

Figure 7.18. Characters are expressed hierarchically based on our knowledge and hypotheses of homology. Homologs are tested and confirmed by congruence with other homologs or are rejected by incongruence.

The areagram includes two more informative three-item statements, E(BC) and D(BC), that are not independent; they are logically implied given the four independent statements. Together, they do not convey enough information to recover the areagram (see Nelson and Ladiges, 1991b): the two dependent statements E(BC) and D(BC) give DE(BC). The fractional weight is 4/6 (four independent statements out of six possible statements; Figure 7.19). The areagram ((AB)C)DE has seven independent three-item statements; this equals the total number of statements that recover the areagram. The fractional weight, 7/7, equals 1.

Unlike a parsimony method that chooses between the general areagrams based on tree length, our goal is to find the *minimal tree* based on information content.

The Minimal Tree

The minimal tree expresses all congruent relationships found in a series of areagrams or subtrees (see Nelson and Ladiges, 1996; Kitching et al., 1998). It is *not* a consensus, in that there is no attempt to resolve

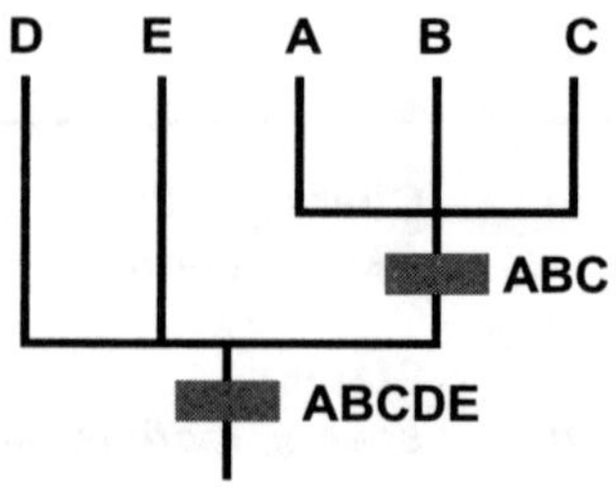

6 informative three-area statements:

independent	**dependent**
D(AB)	E(BC)
D(AC)	D(BC)
E(AC)	
E(AB)	

4 independent statements, added together, recovers areagram DE(ABC).

2 additional, dependent three-area statements recover areagram DE(BC).

Fractional weight = 4 independent statements out of a total of 6 = 4/6.

Figure 7.19. Fractional weighting applied to compare the number of independent statements to the number of overall statements.

partially incongruent areagrams or subtrees. Minimal trees are expressions of relationship and in systematic biogeography represent the general areagram—the branching diagram that expresses geographical congruence, *not* incongruence.

There are different ways to find the minimal tree using different criteria (parsimony, compatibility, etc.). In each case, the minimal tree contains the same relationships found in the input subtree and areagrams. Minimal trees depict general area relationships.

Three-Item Analysis and Compatibility

Systematic biogeography is not a way to justify or explain incongruence. Incongruence may have many explanations, such as extinction,

Box 7.6 Consensus and Minimal Trees

Consensus and minimal trees are used to summarize the relationships between data.

The difference between minimal and consensus trees is that the former searches for *congruent* patterns, whereas the latter seeks to resolve conflict within more than one areagram.

The following two seemingly "conflicting" Pacific patterns are solved differently by consensus and minimal tree methods:

New Caledonia(South America(Australia, New Zealand))
South America(Australia(New Zealand, New Caledonia))

The consensus method finds an unresolved tree, whereas the minimal tree finds the pattern present in both trees:

South America(Australia, New Zealand)

Minimal trees were originally intended to serve as the basic relationships present in single cladograms and areagrams, rather than as a representation of the least number of transformations between character states (Nelson and Ladiges, 1996). Minimal trees are found using compatibility (as above) or parsimony methods. In biogeography, the minimal areagram contains no MASTs or geographic paralogy and is synonymous with a subtree.

failure of a taxon to respond to a vicariance event that caused differentiation in other taxa, and so on. These explanations need to be evaluated taxon by taxon but cannot be explored until the relative degree of congruence is assessed. By minimizing incongruence, geographical congruence is uncovered, which is the aim of a comparative biogeography. *Nelson05* uses compatibility to choose among possible area relationships (see Nelson, 1979; Siebert, 1992; Meacham, 1981; Meacham and Estabrook, 1985). Relationships conflict where there is no congruence; the program minimizes conflict to discover pattern.

DISCOVERING GEOGRAPHICAL CONGRUENCE

Geographical congruence may be used to build biogeographical classifications. As a practical example, we consider taxa that live in the Mexican Transition Zone, a complex area that includes the southwestern United

Figure 7.20. Fourteen biogeographic areas in the Mexican Transition Zone (modified from Escalante et al., 2007: Figure 1). Two additional areas in the analysis are the Nearctic (north of Mexico) and Neotropical (south of Mexico).

States, Mexico, and northern Central America. It is so called because it is home to a biota that includes members of both the classic northern Nearctic and southern Neotropical biogeographic regions of Sclater (see Figure 2.4). Distributional history of taxa throughout this region has traditionally been explained as the Great American Biotic Interchange, following American mammalogist George Gaylord Simpson (1940, 1950): the multiple, overlapping distributions in the transition zone were explained by dispersal from the north of Nearctic elements and dispersal from the south of Neotropical elements. This explanation was readily accepted for formation of the Mexican Transition Zone biota and has rarely been challenged.

A phylogenetic analysis of 21 species of daisy genus *Montanoa,* family Asteraceae, that live in the Mexican Transition Zone resulted in a fairly well-resolved hypothesis (Plovanich and Panero, 2004): ((((((AB)C)(DE) FG)H)(((IJ)K)L))((S((OP)(MN)(QR)))(TU))). Each letter represents one species. Sixteen areas occupied by Mexican Transition Zone species were described by Escalante et al. (2007; Figure 7.20). *Montanoa* lives in 11 of those areas. Most species of *Montanoa* are widespread—that is, they live in more than one area—and one species, K, lives in eight areas.[5]

The well-resolved *Montanoa* taxon cladogram was converted into a MAST-riddled areagram. The potential complexity of area relationships expressed by 21 species living in 11 areas is staggering: over 1 million areagrams resulted when *Nelson05* was applied to extract informative

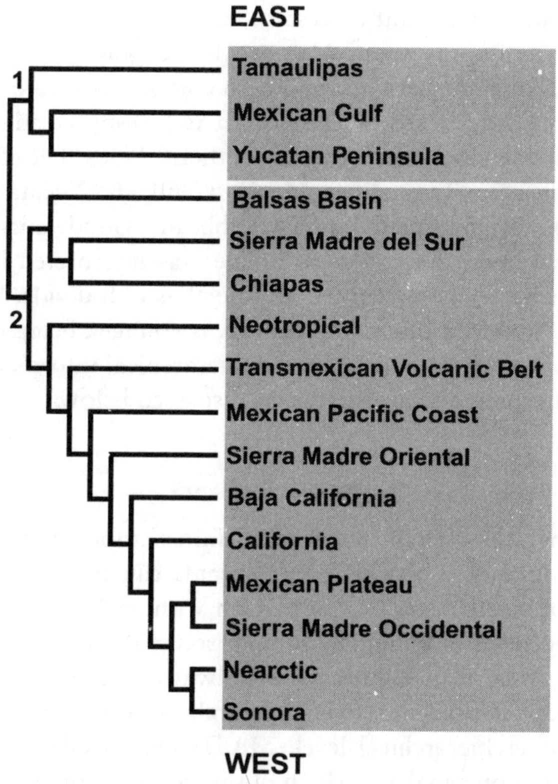

Figure 7.21. Areagram for the 14 biogeographic areas in the Mexican Transition Zone, plus the Nearctic and Neotropical areas (following Escalante et al., 2007: Figure 2; see Figure 7.20 above). Clade 1 (east) and Clade 2 (west) are in turn divided into northern and southern components.

three-area statements (see Escalante et al., 2007). Added together, they produce an uninformative areagram; that is, the *Montanoa* areagram contains no informative nodes despite the well-resolved phylogenetic hypothesis. To extract information on area relationships from the *Montanoa* areagram, the areas need to be redefined to reflect areas occupied by monophyletic groups. How and where this is appropriate throughout the Mexican Transition Zone could be evaluated by considering other taxa that live in the zone along with some potential geological and geographical area limits.

Forty areagrams, including *Montanoa,* of Mexican Transition Zone taxa were examined by Escalante et al. (2007), who produced a general areagram for the 16 areas using *Nelson05* (Figure 7.21). They came to a surprising conclusion: the general areagram was divided into two clades, an eastern and a western, not a northern and southern. Long-held assumptions about Mexican biogeography (that the biota was divided into a northern and southern component) were challenged: challenged, but not discarded. The eastern and western clades are themselves each divided into northern and southern clades. The eastern area, comprising the areas Tamaulipas, Mexican Gulf, and Yucatan Peninsula (Figure 7.20), was interpreted as a Caribbean "Gondwanan" tectonic remnant. Therefore, the east–west divide was interpreted logically as older, of Paleocene age (60 mya). The north–south divide, the pattern of the Great American Biotic Interchange, is younger, being of Miocene age. *Hierarchy analysis* is the way such hierarchical information is conveyed and interpreted in areagrams, as discussed below.

Hierarchy Analysis

General areagrams contain hierarchical information on area relationships. The logically subordinated areagrams can be used to evaluate hypotheses of mechanism, by comparison with geological events, or to evaluate hypotheses of timing, by comparison with fossils, molecules, or stratigraphy. Area relationships are not always straightforward or fully resolved. Area relationships at one hierarchical level may conflict with those at another hierarchical level. MASTs may resolve differently in different taxa in one analysis. Much of this conflict can be explained by the hierarchical level at which one is trying to resolve area relationships: position in the hierarchy is a proxy for age; broadly inclusive clades are logically older than smaller, exclusive clades. Relative timing of events is implicit in area hierarchies; this is the essence of hierarchy analysis.

Area congruence has been interpreted as evidence that areas share a history. Area incongruence has been interpreted as evidence that areas do not share a history, or that areas initially considered as one are better treated as composites, or as indicators of patterns of different ages (e.g., Grande, 1985; Humphries and Parenti, 1986, 1999). Area congruence for taxa of different estimated ages has been termed "pseudo-congruence" (e.g., Cunningham and Collins, 1994); likewise, area incongruence for taxa of different estimated ages is termed "pseudo-incongruence" (e.g., Donoghue and Moore, 2003), in both cases because it is assumed that whether patterns are the same or different, if they are of different ages, they must have different causes. To avoid these problems, it has been suggested that biogeographic analyses be conducted on taxa of the same estimated age. This has also been called time-slicing. Although time-slicing has many practical applications, such as to explain distribution patterns and events of particular geological epochs, it should be applied after, not before, a biogeographic pattern is sought.

The general areagram for the Mexican Transition Zone contains several explicit area homologs, such as the eastern clade 1: Tamaulipas (Mexican Gulf, Yucatan Peninsula), or in the western clade 2: Balsas Basin (Sierra Madre del Sur, Chiapas). The areas in each of these homologs are oriented north to south. That is, area homologs in the north–south divide are contained hierarchically within the east–west divide.

Imagine if the older components—Tamaulipas, Mexican Gulf, and Yucatan Peninsula—had been segregated prior to the analysis so that Paleocene taxa were analyzed separately from Miocene. A north–south divide would be recognized for the Paleocene taxa and for the Miocene taxa, and the east–west division would possibly not be recognized at all.

There is another important point here about understanding the hierarchy of area relationships: two areas do not specify an area homolog. *The eastern and western clades of the general areagram are not necessarily sister areas.* They are depicted as sister clades on the general areagram (Figure 7.21) because the study was limited to 14 biogeographic areas in Mexico, plus the Nearctic and Neotropical regions. A third area, which can be almost any other area in the world where the appropriate taxa live, must be added to the analysis to specify a relationship. That is, to understand the relationship between the eastern and western clades, we must go to a higher hierarchical level.

Solving the relationships among areas is dependent upon hierarchical level and the amount of information (cladistic resolution) at each level for each taxon. How we interpret these different patterns may be influenced by the quality and amount of information we have on timing, such as from molecular data or fossils and the stratigraphic record, but those data cannot refute the patterns.[6]

Examples of "pseudo-congruence" (e.g., Taylor et al., 1998; Hunn and Upchurch, 2001) in which the same biogeographic pattern in two or more groups is viewed as being caused by different events because the groups are of different ages, as estimated from fossils, molecular data, or stratigraphy, need to be reconsidered. A fossil does not demonstrate proof of absolute time and place of origin (see discussion in Craw et al., 1999); rather, it demonstrates that a lineage was present at that time and in that place. Temporal data cannot alter the results of the analysis to produce an alternate areagram. A Pleistocene fossil, for example, is no more or less useful than is a recent taxon for identifying a pattern formed in the Mesozoic.

Our goal is to replace the idea that the purpose of biogeographic analysis is to choose between explanations of dispersal versus vicariance with the idea that the purpose is to discover general areagrams. Systematic biogeography proposes area classifications and tests them to discover general patterns of the distribution of life on Earth. We consider further the relationship between area relationships and geology in Chapter 8 and apply the principles of systematic biogeography to the Pacific in Chapter 9.

SUMMARY

- A first step in any biogeographical analysis is the explicit recognition and description of the study area as the overlap of two or more taxic distributions.
- Limits of the study area are the distributional limits of the organisms under study or an inorganic disjunction.
- Areagrams and TACs are obtained directly from phylogenetic analysis of monophyletic groups.
- Taxa of a single analysis may vary in age.
- Areagrams are summaries of area relationships and contain no information about taxon relationships. TACs are phylogenetic trees that contain areas and taxa at the terminal branches and hypothetical taxa and areas at nodes.
- A single areagram is a relationship or area homolog, whereas patterns are represented in general areagrams. Patterns are evidence for historical relationships or area homology.
- Paralogy-free subtree analysis combined with the transparent method is the best way to resolved areagrams that contain geographical paralogy and MASTs.
- Temporal data are secondary in a biogeographical analysis. They cannot be used to alter the relationships within an areagram because area relationships, regardless of age, are derived from taxon relationships.

NOTES

1. Nodes here mean nodes on a cladogram; this is different from the panbiogeographic concept of a node, discussed in Chapter 6.

2. Eliminating geographically paralogous nodes from an areagram can only be considered misleading in a biogeographic analysis if the data on area relationships supplied by the geographically paralogous node are informative. A geographically paralogous node is uninformative in an areagram, by definition. A geographically paralogous node may be informative in a TAC. Thus, the debate should not be about discarding data, but about discovering areagrams that can be added together to form general area classifications *versus* interpreting biogeographic distributions outside of a comparative framework.

3. *Nelson05* may be downloaded from the Web site of LIS (Laboratoire Informatique et Systématique) Université Pierre et Marie Curie-Paris 6: lis.snv.jussieu.fr/newlis. Web sites for other systematics software include www.cladistics.com and mrbayes.csit.fsu.edu.

4. Unlike most methods in systematics and biogeography, three-item analysis is not based on the transformation of homologs (character states). It is therefore pointless to argue for or against three-item analysis based on transformational principles.

5. Twenty-one *Montanoa* species (A through U) and the areas in which they live: A, vol; B, bal mpa; C, mpa sms bal vol; D, smo chi sms mpa vol bal; E, mpl sms vol; F, mpa; G, mpl vol bal sms; H, mpa chi; I, sms; J, chi neo; K, sms sme vol chi mgu mpa mpl neo; L, vol bal; M, neo; N, chi; O, yuc mpa neo; P, neo; Q, neo; R, mgu chi mpa neo; S, neo; T, mpa bal vol sms; U, sms. Area codes for the Mexican Transition Zone taxa are as follows: 1, nea = Nearctic region; 2, neo = Neotropical region; 3, baj = Baja California; 4, bal = Balsas Basin; 5, cal = California; 6, chi = Chiapas; 7, mgu = Mexican Gulf; 8, mpa = Mexican Pacific Coast; 9, apm = Mexican Plateau; 10, sme = Sierra Madre Oriental; 11, smo = Sierra Madre Occidental; 12, sms = Sierra Madre del Sur; 13, son = Sonora; 14, tam = Tamaulipas; 15, vol = Transmexican Volcanic Belt; 16, yuc = Yucatan Peninsula. (See Escalante et al., 2007.)

6. Areagrams need not be pectinate to represent a "temporal" hierarchy, as Zaragüeta et al. (2004) argued for cladograms, although we agree that when not pectinate, they may imply multiple temporal hierarchies for both taxon and area cladograms (see Chapter 6).

FURTHER READING

Mickevich, M. F., and N. I. Platnick. 1989. On the information content of classification. *Cladistics,* 5, 33–47.

Nelson, G. 1996. Nullius in verba. *Journal of Comparative Biology,* 1(3/4), 141–152.

Nelson, G., and N. I. Platnick. 1981. *Systematics and biogeography, cladistics and vicariance.* Columbia University Press, New York.

Swofford, D.L. 1991. When are phylogeny estimates from molecular and morphological data incongruent? In M. M. Miyamoto and J. Cracraft (eds.), *Phylogenetic analysis of DNA sequences* (pp. 295–333). Oxford University Press, New York.

PART III

IMPLEMENTATION

EIGHT

GEOLOGY AND COMPARATIVE BIOGEOGRAPHY

A BIOGEOGRAPHER'S GUIDE TO GEOLOGY

Biogeographers hail from diverse backgrounds, including geography, taxonomy, systematics, and ecology. Most biogeographers likely align themselves with ecology, the study of the interactions of life, usually at the level of populations or species. Most biogeographers think of evolution (change over time) as taxic or molecular evolution resulting from mechanisms such as adaptation, predation, or competition. Geology, for many biogeographers, is little more than the theater in which the biotic evolutionary drama is played out. Few address the interactive relationship of biology and geology other than to acknowledge brief details of continental drift or more immediate geological phenomena, such as mountain

Overview

Geology is vital for delimiting biotic areas and interpreting the mechanisms that may have formed biological patterns. Yet most biologists have little understanding of geological materials and processes. Here we introduce the physical processes that create different types of rock and discuss how rocks are weathered and eroded, the process of fossilization, and the effects of rock formation on the living world.

Comparing biotic patterns with geological materials and processes leads to a *reciprocal illumination* between biogeography and Earth sciences. Biological patterns do not test geological patterns. Geological patterns do not test biological patterns. One informs the other. Geological mechanisms have been the driving force behind many biogeographic patterns. Biological mechanisms may also be the driving force behind some geological patterns.

Biogeography and geology of the Pacific is introduced as a case study that we examine in more detail in Chapter 9.

building or stream capture. The role of geology in evolution, therefore, is often dismissed: it is "too slow," potential barriers, such as oceans and mountain ranges, "too big." Paradoxically, many biogeographers accept geological explanations, including timing of geological events and paleogeographic reconstructions, without question, and use that information to confirm or reject hypotheses about distributional mechanisms.

Geology is asked to explain long-past events in the midst of "deep time" that may or may not affect present distributions. At the same time, geology is asked to play arbiter among biogeographic mechanisms. Understanding and appreciating how geology and biogeography are interrelated allows a biogeographer to define more meaningful areas of endemism (see Chapter 3), to know where and how fossils are formed, and to propose new geological hypotheses for geologists to test. It also allows greater understanding of biogeographical mechanisms, such as ecological stranding (Chapter 5), that may change the ecology of an entire biota.

Geology is a highly dynamic field that explores a rapidly changing Earth. Topics of geological studies range from earthquakes and the resulting tsunamis that wash away large swaths of land and change the biotic composition of areas to the erosion of rocks and the soil types that result in environments hostile to particular taxa. Understanding geology means understanding one of its most fundamental components: the rock.

To ask what a rock is in geology is akin to asking what a taxon is in biology. Rocks, like taxa, can be classified and divided into subunits. Rocks and taxa can be described based on unique characteristics. Both also have developmental stages, which bear witness to the growth of geological form both in minerals and in rocks. Overall, geology and biology have parallel aims—to diagnose and classify rocks or taxa and to describe their development and evolution over time.

Introducing Geology

Rocks are made of minerals. A mineral has a unique structure based on its chemical composition (analogous to DNA) as well as on its observed traits or morphology. Minerals can be classified into groups based on either their physical structure or their chemical composition. Rocks, too, may be classified using qualitative traits such as porosity and structure, mineral content, or chemical composition.

Rocks may change into other rocks with an increase in heat, pressure, or both. Limestone exposed to extreme heat changes into marble. Shale, when compressed, heats up and deforms into slate; further deformation turns it into phyllite. Rocks that are deformed through heat or pressure are termed *metamorphic rocks*. The slate on your roof and the marble Michelangelo sculpted are metamorphic and are derived from shale and limestone, respectively.

Shale and limestone are made from other rocks, minerals, or organic sediment. Shale may comprise organic detritus and fine sediment that has been eroded from another rock. Limestone is formed largely by the products of reef-building plants and animals such as algae and corals; a coral extracts dissolved salts from seawater that crystallize into carbonates. Shale and limestone are *sedimentary rocks*. Metamorphic rocks that are exposed to weathering and erosion can form sedimentary rocks if compressed, which forces the finer sediment to cement the rock together. Sedimentary rocks may also be derived from rocks formed from lava or magma. Basalt is a rock that forms from rapidly cooled lava. The lava that is extruded from a volcano may cool at different rates, forming crystals of different sizes: slow-cooling lava forms larger crystals, whereas fast-cooling lava will form rocks such as obsidian that resembles dark glass. Basalt and granite, rocks that originate from molten magma that solidified either on (in the case of basalt) or within (in the case of granite) the Earth's crust, are *igneous rocks.*

Most rocks on Earth start off as igneous rocks emplaced either inside the Earth (as *intrusive rock*) or on its surface (as *extrusive rock*). A rock may exist for a long time, changing from one form into another and, in turn, changing the environment in which it forms or erodes. That rock may now exist as a part of other rocks or as a simple pebble washed up onto a beach.

THE JOURNEY OF A PEBBLE: EXPLORING GEOLOGICAL CONCEPTS

When we find a pebble,[1] say in a stream or on a shoreline, we may note that it is smooth and cold to the touch. Perhaps we skim it across the surface of the water to see how many times we can make it skip. Our pebble contains a story of how our Earth works and how a piece of rock may have traveled great distances, perhaps through eroding river banks, to become part of a larger conglomerate rock, perhaps washed out to sea or even serving as a gizzard stone in a bird. Our pebble's story begins in a large body of molten rock that has intruded into the crust of the Earth.

Rock Formation

The Earth's crust is a thin, delicate layer, divided into two shells, covering a large, extremely hot ball of viscous, convecting rock kept solid by the great confining pressure that surrounds an inner sphere of hot iron. The upper crust is a 10- to 12-km-thick layer of continental rock, light in composition and diverse in minerals. The lower crust is dense, relatively thicker (15–20 km), and similar in composition worldwide; it includes oceanic rock. Few humans have seen oceanic rock where it is formed: deep under water, where two plates diverge. The divergence points are *mid-ocean ridges,* possibly the highest mountains on Earth; they extrude basalts into cold water, and these basalts form pillow lavas and more oceanic crust. The cycle is endless, as more and more lava is extruded as the plates are pulled apart, moving them in opposite directions. Unlike oceanic crust, continental crust is much older—some of it is the oldest rock on Earth. There is no oceanic rock older than 170 million years, whereas continental rock can be older than 4 billion years. Continental rock is the rock that we are familiar with; it exists mostly above sea level and consists largely of granite—an igneous rock—along with metamorphic and sedimentary rock.

Igneous rock is extruded from volcanoes either as lava or as ash, or it forms as intrusive rock that enters the Earth's crust as a large body of molten rock, a *batholith*. The majority of volcanoes result from batholiths that extrude molten material through fissures in the rock. Rocks that form in the batholith cool over a long period of time. Batholiths themselves turn into rock and become exposed over long periods of erosion. Molten rock that is extruded from volcanoes as lava forms quickly. You can readily tell the difference between an intrusive and an extrusive rock: both have interlocking crystals, but intrusive rocks that have cooled over longer periods of time have developed larger crystals, as is the case with granite. Extrusive rocks have much smaller crystals; some such rocks, like obsidian, cool so quickly that they resemble glass. Our pebble underwent a brief cooling in a granite batholith (~ 700–800° C) and was extruded gradually in a violent explosion of a rhyolite volcano, such as the now dormant Taupo Volcano in New Zealand. Extruded molten rock is termed lava; intrusive molten rock is magma. Our rhyolite lava, typically low in iron, consists of a series of larger crystals, or *phenocrysts,* of quartz, formed during its time as magma, as well as smaller crystals that formed when it was finally extruded as lava—a process that resulted in the rock having a light salmon-pink color.

From Rock to Pebble

Our pebble was once part of a large formation. A formation is a unit of rock that formed over a particular time period. A volcano may extrude a lava flow that covers an area and hardens to a certain thickness. This can be a series of continuous events that stop abruptly. The result is a formation, which geologists name and assign an age, based on the minerals formed within the rock and an absolute measurement of its thickness. A formation is made up of smaller units called *beds,* which represent events. One eruption is notably different from another because of the presence or absence of water or because of changes in the fluidity of the lava. Fluid lava flows smoothly and forms *flow structures;* the minerals are aligned in the direction of the flow. Where the lava is viscous, the resulting rock contains holes formed via small gas bubbles in the flow. A formation may include thousands of bedding structures,

Figure 8.1. Horizontal bedding, at Santa Barbara, California, Pacific coast of North America. Approximately 1 meter wide. [Photograph by Lynne R. Parenti.]

each of which represents a different deposition event (Figure 8.1). The structures within each rock formation vary in size; they range from hundred-meter-high faults to microscopic folds. The study and classification of formations and their structures is called *stratigraphy.*

The lithology of an area changes constantly because of pressure exerted by rocks lying above or because of erosion of rocks occurring at the surface. *Erosion,* the breaking down of the chemical structure of minerals in a rock or of the rock itself, results from *weathering.*

Our rhyolite has been subjected to many different forms of weathering. Its pink has faded to a light gray. Soil covering the rock contains roots from the vegetation growing above, which aid *mechanical weathering:* roots find their way into small cracks and crevices and slowly pull the rock apart as they grow. The chemical composition of the soil includes compounds that react with the rock and slowly turn it into grayish clay. *Chemical weathering* also results. Rain, as well as hot sun, can change the chemical composition of minerals. Dew and cold evenings cause frost to form in exposed crevices and split the rock open, inviting more soil and run-off. Mechanical weathering breaks the rock down, and chemical weathering breaks the chemicals

Figure 8.2. Examples of fast-moving stream with boulders (above) and slow-moving stream against a vertical section of strata (below), Sarawak, northwestern Borneo. [Photographs by Lynne R. Parenti.]

down into smaller parts and into different compounds, such as clay. Our pebble has been affected by both types of weathering. The roots of a tree pulled apart several large pieces of highly weathered rhyolite caked in soil and clay. After the tree died and fell, the pieces of rock weathered further, becoming smaller. The rock was most likely covered by clay and soil until the day a landslide during a storm deposited it into a gully. The storm washed away most of the rubble on the hillside leaving bare rock.

Most broken-down rock that does not form soil ends up in streams or other fast-moving waterways. Over time, the stream slowly erodes the rock, etching its surface away as the rock gets battered and beaten against the rubble (Figure 8.2). Our rock is no longer part of a formation. It has been eroded and lies detached, with other rubble, away from its area of formation. Our rock has become a form of sediment: a pebble.

Soft Rocks

Geological terminology differs for rocks that are formed as igneous or metamorphic rocks and those that are formed as sedimentary rock. The former are hard rocks, the latter soft rocks. Sedimentary rocks are formed from *sediment,* which originates from rocks, minerals, and occasionally organic residue (e.g., bones, vegetative matter). Sediment size may vary; it can range from microscopic sand grains to large limestone blocks up to a kilometer in length. Geologists can identify the rock and its provenance based on a few simple rules. Through the field of *petrology,* a geologist can classify rocks based on their structure and composition. Through *mineralogy,* rocks are identified simply by identifying the types of minerals present and whether they have been deformed. Using *geochemical analysis,* geologists can identify rocks through their chemical signature.

Once a rock has been identified and named (e.g., basalt), it needs to be placed geographically. Some rocks have such a unique chemical or age signature that geochemists can place them with geographic source areas. Generally, geologists rely on knowledge of the area where the rock was found to identify it and its formation. A rock is identified as autochthonous or allochthonous. Rocks or boulders that were formed in their present position are *autochthonous.* Boulders, pebbles, or other sediments that have moved away from the area where they were formed are *allochthonous.* Examples abound of allochthonous boulders that may have traveled surprisingly large distances. In the Devonian Nubrigyn Member (Formation) near Wellington, New South Wales, Australia, large limestone boulders, usually formed in shallow marine environments, lie in muddy sedimentary rock formed in deep water (Conaghan et al., 1976). For pebbles, the distance from the source rock can be

Box 8.1 Earth and Life and Life and Earth (Reprise)

An organic Earth, in which biological and geological processes influence each other, is a foreign, even contradictory concept to many biologists and geologists. The prevailing idea that the inorganic (Earth) can change the organic (life), but that the reverse is impossible, is contradicted by many local and global examples: Limestone reefs alter oceanic currents, form rock, and create soft, granular soils. Phytoplankton provides an oxygen-rich environment in which minerals form and oxidize. Heat created through forests alters our climate. Organic soils may form rock like bauxite or erode rock and alter water chemistry.

The Gaia Hypothesis, also known as Earth Systems, proposes that Earth is a living organism in which the organic and inorganic parts influence and are dependent upon each other. The idea, proposed by British scientist James Lovelock (1979), echoes Croizat's tenet that life and Earth evolve together, but differs fundamentally: the Gaia Hypothesis allows for life to *alter* Earth systems. As one example, geologist Don L. Anderson proposed that biomineralization in the form of calcium carbonate or limestone deposits may be responsible for cooling the Earth's surface and upper mantle and that "there is the interesting possibility that plate tectonics may exist on earth because limestone-generating life evolved there" (Anderson, 1984:348). That is, life created the conditions under which plate tectonics could occur.

Biology is so intricately linked with geology that the ages and periods of Earth (e.g., Mesozoic) are marked by life's milestones, including the diversity and composition of fossil biota. That life and Earth each affect the evolution of the other is undeniable.

estimated by the degree of roundness. Pebbles usually lie in streams or rivers, and the degree of roundness depends on the amount of abrasion that the pebble has received. An angular pebble is most likely to be closer to the source rock, whereas a smooth, rounded pebble will be farther away, having been exposed to much more weathering. The amount of weathering and erosion of a pebble depends on how far it has moved. Sediment in rivers constantly moves: finer-grained sediment moves more quickly because it has little mass, whereas gravel, pebbles, and more massive boulders move slowly.

Water velocity also determines the sort of sediment transported. A fast-flowing river transports fine and some coarse-grained sediment, whereas a slow-moving river moves barely any large grains. A fast-moving mountain stream will have washed out most of its fine sediment onto banks, leaving the larger sediment, whereas a slower moving river on a flood plain will be turbid and will have muddy banks. Our pebble moved within a swift river, gradually becoming more rounded as it was taken farther away from its source rock.

Becoming Rock

Sediment may undergo compaction, and sometimes cementation, through exposure to pressure and heat, eventually becoming sedimentary rock. The process of sedimentation can be observed while walking along a beach, reef, river, or sand dune. When fine grains of sand or larger gravel or pebbles rest on a sand bank, beach berm, or reef lagoon, they invariably undergo sedimentation. Layer builds upon layer, much the way a shell is buried slowly wave after wave, each of which deposits sediment.

Sedimentation can take up to a few hours or days on a beach or be instantaneous following a flood, landside, or underwater mudslide. Beaches are places of high sedimentation. Rocks that form on beaches have thin layers (or beds) because the sediment is first eroded somewhat, and then new sediment is deposited. The time intervals between each layer can vary greatly, as beaches can be eroded away entirely in short periods of time, changing the ecosystem (from sandy beach to rocky shoreline) within a few hours. Where sedimentation is high, such as in a flood, the bedding is thicker and will survive further erosion. An increase in sedimentation leads to compaction of lower-lying beds.

Compaction results from overlying pressure exerted by increasing sedimentation; this pressure forces the sediment together. Minimal compaction results in a porous and brittle rock, such as reef rock or *coquina,* common on Florida beaches. Compaction rates vary depending on how the sediment is sorted (whether as equally or unequally sized grains) and on the quantity of cement present (e.g., calcareous mud). A muddy rock, such as reef rock, which contains unevenly sorted grains, will solidify quickly, whereas evenly sorted grains with little pore space, such as dune sand, will take longer to lithify. An increase in pressure leads to an increase in temperature. The chemical composition of finer grains alters under temperature, and they fuse together, forming a hard mortar or *matrix.* Once this occurs, the sediment becomes rock: either fragile coquina or harder sandstone. Lithification, the formation of stone from loose fragments, grades into metamorphism.

Our pebble, deposited on a sand bank during a storm, is covered by finer-grained sediment as the river upstream cuts into a softer mudstone. The river slows and the turbidity increases allowing thicker and denser layers of sediment to settle upon the unevenly sorted medley of grains. The finer grains from the mudstone slowly glue the larger grains together. Heat from the overlying pressure of sediment buries the slowly forming rock beneath other rock, sediment, and soil. The river above has vanished, and a prairie now lies in its place. Our pebble has become part of a conglomerate rock. If pressure and heat were to increase, the conglomerate would slowly change. Minerals would change under pressure, react to the heat, and re-form as new compounds. The heat and pressure would also squeeze the grains, pebbles, and finer sediment like plastic, forming new structures that reflect the direction of push and pull, and thus forming a metamorphic rock. Sedimentary rocks preserve the remains of living tissue or fossils, but metamorphism usually destroys these.

Fossilization

Paleontology, the study of past life, deals with fossils or fossil remains. A fossil is *any* naturally preserved organism or its traces, living or extinct. Organisms that have been preserved relatively quickly, such as woolly mammoths or burrowing crustaceans, may form *subfossils.* These include fossils of species that are still extant or have recently become extinct (see Ahyong and Ebach, 1999).

Becoming a fossil is difficult. Organic matter decays astonishingly quickly and is rarely preserved before the start of decomposition. Fossils may be preserved due to larger geological events. The fossil plants of the Sydney Basin, Australia, were preserved during flash floods; the Burgess shale fauna of the Canadian Rockies was possibly preserved when earthquakes caused underwater mudslides. Other fossils have been preserved due to the state of the inorganic environment. The biota of Solnhoffen, Germany, site of *Archaeopteryx,* the iconic fossil bird, was preserved as marine organisms were flushed into a toxic lagoon high in salt; the fossils of Messel Pit, Germany, were possibly gassed or trapped in anoxic bogs and swamps, and Baltic amber insects, reptiles, and amphibians were caught in resin flows. These particular environments and geological events result in unusually high-quality, well-preserved fossil specimens in large numbers, found in strata known collectively as *Fossil-Lagerstätten.*[2] Organisms preserved under such ideal conditions may be exceptionally complete, with preserved soft tissue in some cases, whereas the majority of fossils elsewhere are merely the hard parts of organisms, such as teeth, shells, or bones, which have survived further decomposition.

Given the dynamic nature of life on Earth, anything that dies and is not preserved instantly will generally disappear. Apart from rare cases where organisms are covered in sticky and lethal sap or fall into a tar pit, most life will not be preserved in its entirety.

Large-scale fossilization usually depends on sedimentation or deposition in water. Strictly terrestrial fossils are rare, and those land-dwelling organisms that did become fossilized did so because they lived or died near a river, lake, stream, lagoon, or tar pit. Even amber needs water to preserve fossils. Resin is prone to oxidation and can only survive if it is preserved rapidly. Not surprisingly, most of the life forms preserved in the fossil record are marine organisms. To appreciate how rare fossilization is, we look back 170 million years to the Jurassic.

The Jurassic (200 to 145 mya) is a period in the Middle Mesozoic. Paleontologists have reconstructed the paleoenvironment and atmosphere using fossils and chemical isotopes at various localities throughout the Jurassic world. Once we understand the few habitats where organisms can be preserved, we can appreciate how little of life was preserved. Fossilization generally occurs in water: not fast-flowing rivers or beaches, but slow-moving, turbid, and (ideally) anoxic environments, such as stagnant ponds, slow-flowing muddy streams, or reef lagoons. The majority of such environments are marine: saltwater environments, such as shallow seas, the ocean floor, sheltered reefs, and shorelines. All known marine environments before the Jurassic were on continental plates: no oceanic plate older than 170 million years has been preserved. Today's oceanic plate environments include deep abyssal plains and coral reefs, the most biologically diverse habitats on Earth. During the Jurassic, many continents were flooded, forming marine environments called *epicontinental seas*.

Paleontologists living 170 million years from today will know little of our 21st-century world. A few fossils may be formed of organisms with hard parts such as bones, shells, and exoskeletons, but only the largest, sturdiest, and most common species will be represented. Perhaps a whale bone, layers and layers of eucalypt leaves and fruits, large formations of *Acropora* coral, and a few dung beetles buried in mud will be found, and there will be mollusks, diatoms, and shark teeth galore. But future humans will know little or nothing of nudibranchs,

Box 8.2 How to Fossilize

Most organisms alive today will not fossilize. Perfect conditions for fossilization are rare. Ideally, you would need to be near turbid, almost stagnant water—like a foul-smelling pond, for example—to get a chance at fossilization. Even if you do fossilize, you may not last: many strata are soon eroded away.

Here is a guide to fossilization in seven steps.

1. Have at least some hard parts: bones or an exoskeleton are best. You are likely to be preserved if you have lots of waxy leaves or the specialized, tough skin of a pollen grain.
2. Find a lagoon that is not prone to much tidal activity or that does not wash in too much coarse sand. Mud and silt are the best materials for preservation; sand is coarse and abrasive.
3. Sedimentation should be constant. Pick a lagoon or bay.
4. The water should have an anoxic layer. Deeper bays or lagoons have "dead" zones where there is little oxygen. Oxygen just attracts things that will eat you, such as fish, worms, or protists, before you fossilize.
5. Die in or near the water so there is a good chance that you will sink. Decomposing organisms tend to float. This is problematic, as you will be fed upon by fish and larger prey (see 4). Dying in freshwater decreases buoyancy and the chance of being eaten by larger organisms.
6. Make sure you are far away from mountains that are likely to have rocky rivers that may erode your final resting place.

bees, flies, freshwater fish and crustaceans, hummingbirds, roses, cacti, or grasses. Even if an insect were well-preserved to start with, it might not withstand the constant battering and grinding of weathering or the effects of metamorphism. Erosion and weathering are constant. Fossilization is a privilege. Erosional regimes, such as exposed bedrock, will yield no fossils, whereas depositional regimes may.

Fossilization preserves not only the organism, but details of the fossilization process and the environment the organism lived in as well as some aspects of the organism's life. *Teratology* is the study of

Box 8.3 Geology and Biogeography

Figure 8.3. Barnes Butte, Papago Park, Phoenix, Arizona. [Photograph by Malte C. Ebach.]

Barnes Butte is made of a red, mid-Cenozoic breccia, a sedimentary rock, which overlies pre-Cambrian granite in Papago Park, Phoenix, Arizona (Campbell, 1999). The stark contrast between the bare hill, devoid of plant life, except for a few lichens, and the diverse biota below it on the breccia, illustrates the critical role geology plays in biogeography.

Granites here produce a hard clay soil in which plants can become established and grow. These soils are rarely washed away during the late summer monsoon and are able to support large barrel cacti *(Ferrocactus wislizeni)*, creosote bushes *(Larria tridentata)*, and brittlebush *(Encelia farinosa)*. Granites consist of minerals that are a source of nutrients and often have algae growing on boulder surfaces.

The breccia contains large clasts of granite in calcium carbonate cement and forms a porous, brittle rock. The rock eventually weathers and erodes, forming gravel and sand, which may wash away, leaving no support or nutrients for plants. Two different, yet adjacent, lithologies may support two different biotas.

deformities in specimens; *taphonomy* is the study of the process of organism decay before and during fossilization; *paleoichnology* is the study of fossil traces, such as footprints, tracks, and burrows. *Paleobiology,* a term coined by Othenio Abel (1912), covers the evolution and behavior of organisms based on fossil evidence: fossil eggs and nests inform us about reproduction, embryology, and group behavior, whereas coprolites and gizzard stones tell us about diet.

Without an active and rapidly evolving inorganic Earth—one that weathers, erodes, and forms rocks, rivers, seas, and fossils—life would not be sustained or preserved.

SYSTEMATIC BIOGEOGRAPHY AND GEOLOGY: RECIPROCAL ILLUMINATION

Often the biogeographer is faced with a dilemma. The study area lacks the geological evidence necessary to make a hypothesis regarding the distribution of the biota: there is no evidence of inferred barriers or traces or inferences of other geographical or geological features, processes, or interactions. The problem is acute in marine areas. The dynamic nature of marine areas and their geology and oceanography mean that little or no evidence is preserved to suggest which geological events may have been responsible for the current distribution of biota. Considering that geological evidence (e.g., formation of a rift lake) can be compelling, many rely on biological processes such as dispersal when such geological evidence is absent. One classic example concerns the Pacific Ocean.

Our understanding of the geological evolution of the Pacific Ocean is vague. The majority of the geological evidence that may illuminate its history is lost. No rocks are older than 170 million years, and few of these preserve any history of the ocean's earlier geology. When geologists tell us of the history of the Pacific Ocean, they often rely on what has happened elsewhere, such as in the Atlantic or Indian oceans. As with any science of retrodiction, evidence that supports history of one area does not necessarily support the same history of another—to assume so is to be non-empirical. The early history of the Pacific Ocean will remain a mystery if we rely solely on geological evidence (paleomagnetics). There is at least one way to hypothesize what geological barriers may have been effective: we can study the modern biota. Springer (1982; Chapter 3) identified the Pacific Plate as an area of endemism for marine shore fishes and other taxa by mapping recent distributions.

Box 8.4 You Say *Gondwanaland,* I Say *Gondwana*

Gondwana and *Gondwanaland,* used interchangeably, refer to the same supercontinent, which existed in the Late Paleozoic and Early Mesozoic, over 250 million years ago. It was an amalgam of present-day Africa, South America, Australia, India, Madagascar, and Antarctica.

The term *Gondwana* was likely coined in 1873 by Henry Benedict Medlicott, a 19th-century geologist who used it to refer to the Satpura Basin in Madhya Pradesh (Ghosh, 2002) in northern central India. Gondwana is from the Sanskrit *Gondvana,* referring to *Gonda*—the name of the Dravidian people from an ancient region or kingdom located in present-day northern central India—and *vana,* meaning forest.

Possibly to avoid confusion with the geological formation in northern central India, a site of *Glossopteris,* Eduard Suess (1885) coined the term *Gondwanaland* to refer only to the supercontinent (see Ghosh, 2002). Confusion between the supercontinent and the geological formation is avoided if one follows current geological classification or refers to any Late Paleozoic geological basin in northern central India as a *Gondwanan Basin.* Hence, we use Gondwana.

The patterns of life result from a long association and interaction with the changing geology and landscape of our planet. Inferred barriers that were in place 5 or 120 million years ago are still reflected in the patterns of today's biota. Organisms can, from time to time, ignore such impediments and overcome geographical or geological topographies or compositions. Many biotas have strange distributions and do not reflect current topographies or other chemical, climatic, or physical properties that we may think of as barriers. Regardless of the magnitude of current physical, chemical, or climatic boundaries, they may not be those that have formed the distribution of the current biota. The Pacific Ocean appears to be a barrier to some taxa, but not to others. Even though thousands of kilometers of ocean separate the only two species of the "coral plant" genus *Berberidopsis* in Chile and Australia, they are classified in the same biotic area: the Austral region (Moreira-Muñoz, 2007). Disjunct distributions, even those across desolate areas, do not specify any distributional mechanism.

One way of achieving the goal of reciprocal illumination is to use the relationships of broadly distributed biota to reconstruct the past boundaries of physical areas (e.g., continental margins, mountain ranges, or climatic barriers) or areas defined by chemical composition (e.g., soil chemistry, water acidity, or turbidity). Another way is to use geological history to interpret the general areagram. Geological history cannot reject our patterns of area relationship based on taxa, but it can inform us of possible mechanisms that formed that pattern. A third, yet little realized, goal of reciprocal illumination is to interpret how life may have affected geology (see Box 8.1).

Reciprocal Illumination and the Pacific Ocean

Some of the most compelling fossil evidence for the existence of Gondwana came from the Permian gymnosperm *Glossopteris*, found on all the southern continents. The distribution links the continents together, but in what configuration? The current reconstruction of Gondwana follows that of the South African geologist Alexander du Toit, rather than that of German meteorologist and initial proposer of continental drift, Alfred Wegener (see Chapter 1; Tarling, 1972). Du Toit (1937) attempted to match the geographical boundaries of continental coastlines using geometry and geological evidence. This was significant for two reasons: the coastlines matched without any reference to continental margins, unknown until the mid-20th century, and there was little or no evidence linking India with Australia apart from Permian, Triassic, and Jurassic paleobotanical similarities. The first problem was overcome in the 1970s, when geometric fit was tested against geological evidence (see Smith and Hallam, 1970). By this time, the theory of plate tectonics was being developed, and models of former plate margins were constructed using the direction and speed of seafloor spreading. The connection between areas, such as India and Australia, which had no obvious geometrical connection and little or no shared geological evidence, were modeled based on the fit between other continents (e.g., that between South America and Africa, or Australia and Antarctica). Moreover, the relative speed at which India was moving during the break-up of Gondwana was based on paleomagnetic data.

The problem of the India–Australia connection is still relevant. There is more biogeographical evidence to connect Australasia with South America, counter to the geological reconstruction which links with India and Africa. Such contradictions may characterize geological and biogeographical evidence. The biogeographer needs to evaluate which evidence is stronger or more empirical to best judge which data should be used for geological reconstructions. As for the India–Australia connection, geological evidence is based mostly on modeling, rather than on empirical evidence, such as shared geological formations or common geological structures, such as continuous belts. When biogeographical

Box 8.5 Gondwana A and B

Connection between the west coast of Africa and the east coast of South America has been proposed since at least the 16th century. What confused naturalists was not the geometrical fit of the coastlines (other configurations are possible; see Dobson, 1992: Figure 1), but similarity of the South American and Africa biota. How can two different places share similar biota? Late 18th-century naturalists such as Buffon suggested that there had been some physical connection between the two continents. Considering that there were no islands to which organisms could hop to travel between the two continents, only one solution was possible: the continents must have been connected or at least must have been much closer together. Geographical fit supported the proposal of a large, single continent; this was later corroborated by geological evidence, including seafloor spreading, confirming the hypothesis of continental drift.

A new generation of scientists has questioned the use of one form of evidence—the connection between Africa and South America—to justify connections between other continents (Carey, 1976, 1988; Owen, 1976; Nur and Ben-Avraham, 1977; Shields, 1996; McCarthy, 2005). Given that the size and shape of Earth requires it to be roughly constant through the Phanerozoic (the Paleozoic, Mesozoic, and Cenozoic eras), evidence that supports a South American–African–Indian configuration (Gondwana A) will contradict an Australasian–South American version (Gondwana B). Evidence for an Indian–Australasian connection is based on geological modeling and paleontological similarities.

Gondwana A is also a product of cultural history. The majority of naturalists and geologists resided in Europe or in North America and viewed the planet from a Eurocentric perspective. Would the configuration of Gondwana A be the same if naturalists in the 19th century had

1. Known the shape of all the plate boundaries?
2. Known of pan-Pacific biotic relationships?
3. Viewed Earth from a different vantage point?

If one views the world as if Port Moresby, Papua New Guinea, rather than Rome, Italy, is its center, then the Pacific Ocean is whole, and the Atlantic is bisected, with its two sections lying on either side of the map. Had naturalists viewed the world from this perspective (considering not just shorelines, but also plate boundaries), they would likely have seen the geometric fit between Australia and South America, and North America and Asia. Moreover, biotic links of these connections are

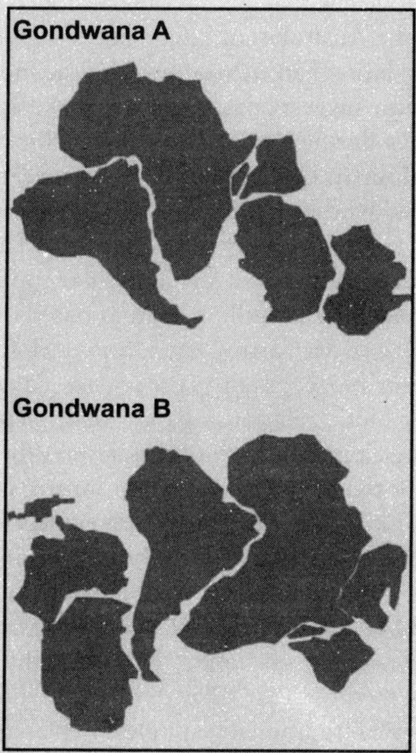

Figure 8.4. Gondwana A: traditional arrangement of Gondwanan continental masses. Gondwana B: alternative arrangement of Gondwanan continental masses, showing relative placement of modern New Guinea, following McCarthy et al. (2007: Figure 4). [Images by Adrian C. Fortino.]

greater than those between the west coast of Africa and east coast of South America: Moreira-Muñoz (2007) reports that 173 genera of plants are shared between New Zealand and South America—an area also known, in part, as the Austral region. McCarthy (2003, 2005) proposed the same relationship in his trans-Pacific zipper effect hypothesis: a closed Pacific opened after the Jurassic, 170 mya. Had the same relationships discovered in the 20th century been recognized by 19th century naturalists, Gondwana would have been configured differently: Gondwana B. Gondwana A and B do not differ in paleolatitude, meaning that paleomagnetic data, the principal form of geological evidenced used to support Gondwana A, also support Gondwana B (McCarthy et al., 2007).

evidence far outweighs that of any other evidence, a compromise may not be best. If the Australasian biota shares a greater affinity with the South American biota than either does with the Indian biota, then reciprocal illumination suggests that Australasia was geographically closer to South America during the formation of the biota than it was to India during the break-up of Gondwana. The biogeographical evidence is further strengthened by the absence of geological formations in the Pacific that date back to the break-up of Gondwana. The only evidence that connects Australasia directly to any particular continent other than Antarctica during the last 250 million years is based on biotic relationships and not geology (see McCarthy, 2003, 2005; McCarthy et al., 2007).

Proposing that biogeography is a science equal to geology is controversial. That biotic relationships are real, plentiful, and able to be discovered using empirical methods underlines the status of biogeography as a state-of-the-art science that can inform geologists of the position of former geological barriers where evidence is missing from the geological record. Wegener relied on biotic distribution data to bolster his proposal of continental drift. Biological hypotheses are thought to be undermined by "harder" geological data or mathematical modeling. Recent developments in systematic biogeography give biogeographers empirical ways to show that their data and hypotheses support plausible alternatives to traditional geological explanations.

SUMMARY

- Rocks are an amalgamation of minerals that are formed in the Earth's upper mantle from other rocks.
- Rocks can be eroded, weathered, and transported as sediment.
- Different types of rocks and minerals can be classified according to different taxonomies.
- Fossilization results from sedimentation and can inform us about paleoenvironments.
- Relationships among biotic areas may inform us of former geographical boundaries.
- Biology and geology have comparable roles to play in our understanding of the evolution of life and Earth.

NOTES

1. The story of a pebble has served as a way to teach the geological history of the Earth in numerous texts at all educational levels. One notable children's book is *The Pebble in My Pocket: A History of Our Earth,* by Hooper and Cody (1996).

2. *Fossil-Lagerstätten* was coined in 1985 by Adolf Seilacher, after a German mining term that denotes a particularly rich seam of ore (see Nudds and Selden, 2008).

FURTHER READING

Christopherson, R. W. 2008. *Geosystems: An introduction to physical geography,* seventh edition. Prentice Hall, Upper Saddle River, New Jersey.

Cutler, A. 2003. *The seashell on the mountaintop: How Nicholas Steno solved an ancient mystery and created a science of the Earth.* Dutton, New York.

Grotzinger, J., T. H. Jordan, F. Press, and R. Siever. 2006. *Understanding Earth,* fifth edition. W. H. Freeman, New York.

Nudds, J. R., and P. A. Selden. 2008. *Fossil ecosystems of North America.* Manson, London.

NINE

IMPLEMENTING PRINCIPLES

Biogeography of the Pacific

THE CHALLENGE OF PACIFIC BIOGEOGRAPHY

Pacific biotic distributions have challenged biogeographers for over two centuries, since the renowned voyages of British naval captain James Cook in the late 18th century discovered an array of startlingly diverse plants and animals unknown to western science.[1] Study of the Pacific biota revealed coherence of life throughout the basin as exemplified by numerous trans-Pacific tracks and sister groups, circum-Pacific distributions, and Pacific endemics. Coherence of a Pacific biota has supported various theories of coherence of Pacific geology, such as "expanding Earth" (e.g., Shields, 1979, 1983, 1991) and the correlated matching of trans-Pacific coastlines (McCarthy, 2003, 2005).

Overview

General patterns of global area relationships are well corroborated, including the general classification of global areas: (Indo-West Pacific(Atlantic, East Pacific))(Boreal, Austral). Support for relationships of areas within these global regions is weaker. We choose the Indo-West Pacific region to implement our comparative biogeographic method. Phylogenetic analyses of taxa in areas of endemism form the raw data of area relationships. They specify area homologs that are combined into an area classification or general pattern. Patterns of the Indo-West Pacific are in turn related to global patterns.

The Indo-West Pacific has long been noted as a center of biodiversity for both marine and terrestrial taxa. It has also been known as the region where distinctly Asian and Australian biotas meet and intermingle. Teasing apart relationships among areas of endemism within this region is facilitated by recognizing and naming areas that are meaningful within a comparative biogeography.

Information on area relationships, as extracted from a series of clades, is represented in an area classification: it incorporates overlapping areas as well as geologically composite areas.

The coherence of Pacific life contrasts with the incoherence of methods and explanatory mechanisms proposed to understand its complexity. The rich and varied literature of Pacific biogeography has been interpreted by Kay (1980) and Springer (1982), in particular, who emphasized that debate in Pacific biogeography has been about mechanisms: dispersal versus vicariance (see Chapter 5). The predominant, traditional biogeographic interpretation is that Pacific oceanic islands received their biota via overseas dispersal from continental source areas. Mechanisms to explain Pacific distributions thus largely focus on long-distance dispersal through the sea (e.g., Ekman, 1953; Briggs, 1995, 1999), migration across land bridges (e.g., Coyle, 1971), or island-hopping by various means (e.g., Gressitt, 1961). Vicariance (e.g., Croizat, 1958; McCoy and Heck, 1976; Nelson and Platnick, 1981; Kay, 1980; Springer, 1982; Heads, 2005b; Renema et al., 2008) is given far less attention.

The debate on mechanisms has been carried out largely without proposals of area homology. To assume that oceanic islands must have been colonized by long-distance dispersal, interpret all data within that framework, and then conclude that long-distance dispersal is the dominant distributional mechanism is perhaps the most pervasive tautology in Pacific biogeography. Likewise, a trans-Pacific sister group pair is insufficient to support a geological model of a former Pacific continent because two taxa or areas cannot specify an area homolog.

The number of species that live in particular areas became a distributional model itself: a high number of species live in and around the center of the Indo-Malay-Philippines Archipelago, and the number of species declines eastward across the Pacific and westward across the Indian Ocean. The drop-off in species numbers is particularly dramatic for some taxa as one crosses the edge of the Pacific oceanic plate (e.g., Springer [1982] for shorefishes). These species and higher-level taxic numbers have been used to support the broadly accepted notion that the Indo-Malay-Philippines Archipelago is a center of origin for marine and freshwater taxa that have evolved in and subsequently dispersed from that center to become established where they live today (e.g., Stehli and Wells, 1971; Briggs, 1999).

Although perhaps compelling, such a mechanism is untestable and relies on a set of arbitrary assumptions, such as the idea that the center of origin must contain the most species, and therefore that the presence of fewer species means one is getting farther away from the center of origin. For some groups, such as *Acropora* corals, a markedly high number of species in the center of the archipelago has been interpreted as a result of the overlap of "broad geographic components to either side of Indonesia" (Wallace, 1997:365). The high number of taxa is interpreted as a result of faunal overlap, not dispersal out of and away from a center.[2] Also, Pacific island and reef size decreases from west to east, which correlates with numbers of taxa but does not support any distributional mechanism. Most important, the center of origin model, by concentrating on numbers of taxa, ignores patterns to be discovered by comparing the phylogenetic relationships among the various taxa that live throughout the Pacific.

To implement our comparative biogeographic method, we identify areas of endemism in and around the Pacific for an array of taxa combined with geological limits and features through time. Phylogenetic analyses of taxa in areas of endemism form the raw data of area relationships. We propose area homologs and combine them into an area classification or general pattern and relate Pacific patterns to global patterns. To begin, we summarize briefly what we know about global areas and their relationships.

GLOBAL ENDEMIC AREAS

The earliest biogeographic maps depicted widespread, global regions. These maps were of terrestrial regions, as marine regions were less well-known biologically and were thought to be ecologically and geologically more uniform than we now appreciate. Twentieth century marine biogeographers were led by Sven Ekman, who reignited the field (see Chapter 2). Ekman divided the global tropical marine biota into four provinces: Indo-West Pacific, East Pacific, Western Atlantic, and Eastern Atlantic. Each province was identified by modern taxa considered relicts of a once widespread Tethys Sea[3] biota dated from at least the Paleogene (65 to 23 mya). The East Pacific fauna was recognized as being more closely related to the Atlantic than either was to the Indo-West Pacific, an area homolog that has been corroborated by decades of investigation on an array of marine taxa. Temperate and cold-water regions of the northern and southern hemispheres were recognized as distinct from the tropics, with their own endemics, demonstrating bipolar or antitropical distributions: taxa in the austral and boreal regions are more closely related to each other than either is to the taxa of the tropics, another well-corroborated global area homolog. We combine these two area homologs, Indo-West Pacific(Atlantic, East Pacific) and Tropical(Boreal, Austral) into a general classification of global areas: (Indo-West Pacific(Atlantic, East Pacific))(Boreal, Austral). This is an areagram, not a taxon-area cladogram (TAC). No optimization of areas is applied; hence, no center of origin is implied.

A global biogeographic classification with boreal (or holarctic), tropical (or holotropical), and austral as the three principal terrestrial areas was proposed formally by Juan J. Morrone (2002; Table 9.1). No relationships among the three areas, ranked as kingdoms, nor among the included regions of each, was proposed or implied. This classification is a consensus or summary of terrestrial areas of endemism on a broad scale. Biogeographers will be familiar with the regions and can readily enumerate their taxa: the Cape honey bee, *Apis mellifera capensis*, is endemic to the Cape or Afrotemperate region, 3.3 (TABLE 9.1); the Australotemperate region is home to numerous endemics in the wet sclerophyll forest of southeastern Australia (see Chapter 3).

A straightforward example will demonstrate the generality of these natural endemic areas. *Trachurus* is a genus of marine fishes known commonly as jack mackerels that lives in tropical, subtropical, and

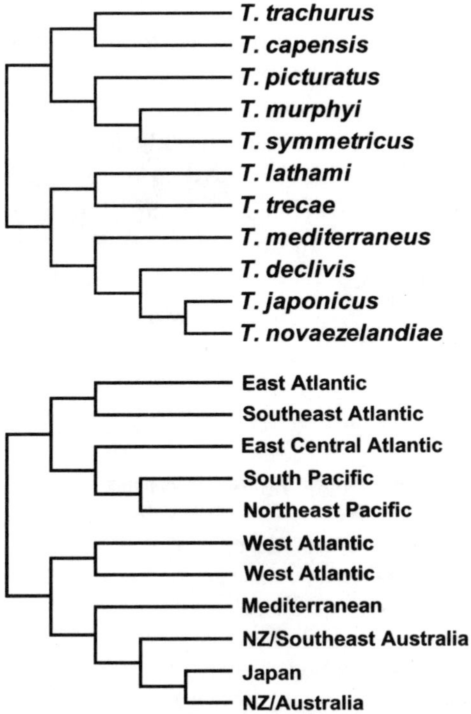

Figure 9.1. (A) Phylogenetic hypothesis of 11 species of jack mackerels, genus *Trachurus*, following Cárdenas et al. (2005). (B) Areagram of *Trachurus*.

temperate coastal and oceanic habitats in all oceans. A phylogeny of 11 of the some 14 recognized species was estimated with molecular data (Cárdenas et al., 2005; Figure 9.1a; Indian Ocean species were unavailable for molecular study). A brief description of the area in which each species lives replaces the name of the species in the areagram (Figure 9.1b). The 11 species are divided into two sister clades, one with five species, the other with six. Each clade has an antitropical sister-group pair: *T. murphyi,* broadly distributed throughout the South Pacific is sister to *T. symmetricus,* endemic to the Northeast Pacific. Likewise, *T. japonicus,* from Japanese waters, is sister to *T. novaezelandiae,* which lives throughout coastal New Zealand and Australia. New Zealand and Australia are geographically paralogous on this areagram: they repeat, and the repetition adds nothing to our understanding of area relationships. Thus, the antitropical regions are sister to Atlantic regions, and the area homolog, as specified by these 11 species of jack mackerel, is

Box 9.1 Area Classification and the Distribution of Anguillid Eels: Why Areagrams Matter

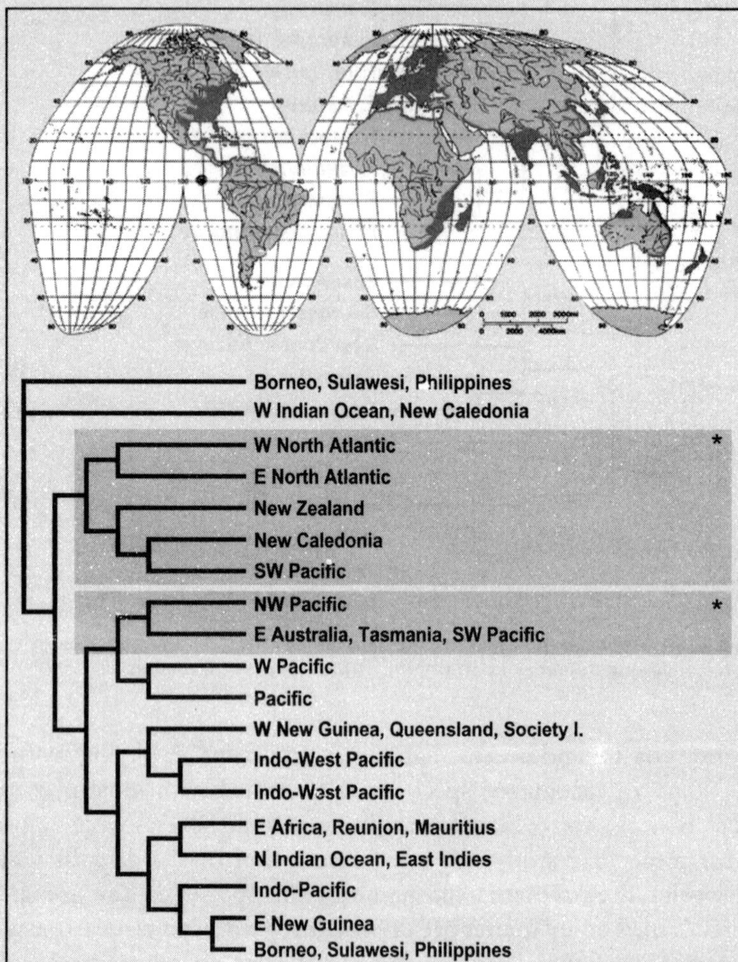

Figure 9.2. Distributional limits of the freshwater eels, family Anguillidae, above (modified from Berra, 2001:69; map copyright Tim Berra), and areagram, below (from Parenti, 2008).

Atlantic(Boreal, Austral). Note that the area homolog, too, repeats, once in each of the two sister clades. Also, antitropicality is focused in the Pacific, not the Atlantic (see Box 9.1). Fossil *Trachurus* from what is now the Mediterranean date from the Early Miocene, some 20 mya.

Freshwater eels of the family Anguillidae are catadromous, a form of diadromy that involves regular movement between freshwater and marine habitats. Anguillid eels enter marine waters to spawn, and offspring return to freshwater streams to complete the life cycle. Anguillid eels have an intriguing distribution pattern: they do not live in the South Atlantic or the Eastern Pacific (map from Berra, 2001). The migratory life history pattern combined with the unique distribution pattern has encouraged unique explanations for the distribution. These explanations have overwhelmingly included dispersal from an inferred center of origin around Borneo, Sulawesi, and the Philippines (see Heads, 2005b). Migration has thus been implicated as the driving force behind anguillid speciation.

Explanations for eel distribution may also be sought in Earth history. Anguillid eels are not distributed randomly throughout the oceans. The relationship between eels and geotectonic features is well known: *Anguilla japonica* makes a spawning migration limited to the margins of the Philippine tectonic plate; the northern Atlantic European *Anguilla anguilla* and North American sister species *A. rostrata* both return to spawning grounds in the Sargasso Sea in the Atlantic Ocean. Earth history is here the driving force behind anguillid speciation.

Alternate explanations for anguillid distribution are also supported by the areagram of anguillid eels (modified from Parenti, 2008: Figure 6). Anguillid eels exhibit remnants of an antitropical distribution in at least two clades, shaded and starred on the areagram. Thus, whatever explanations may be sought for antitropical distribution patterns in general should also include consideration of the eels.

This is the minimum estimate of the age of the group. It agrees with ages of taxa that were once broadly distributed throughout the Cenozoic Tethys Sea.

SYSTEMATIC BIOGEOGRAPHY OF THE PACIFIC

Despite coherence of life around the Pacific, it is not a single, isolated biogeographic region, as Ekman demonstrated readily for the marine biota. Rimmed by a mosaic of terrestrial areas, it includes part of five of the six terrestrial biogeographic regions of Sclater (1858; see Chapter 2) and part of the holarctic, austral, and holotropical kingdoms of Morrone (2002).

TABLE 9.1 RANKED TERRESTRIAL BIOGEOGRAPHIC CLASSIFICATION (MORRONE, 2002)

1. **Holarctic kingdom:** Europe, Asia north of the Himalayas, northern Africa, North America (excluding southern Florida), and Greenland.
   - 1.1. Nearctic region: Canada, most of continental United States, and northern Mexico.
   - 1.2. Palearctic region: Eurasia and Africa north of the Sahara.

2. **Holotropical kingdom:** the tropical areas of the world, between 30° S and 30° N latitudes, including western Australia.
   - 2.1. Neotropical region: tropical South America, Central America, south-central Mexico, the West Indies, and southern Florida.
   - 2.2. Afrotropical region: central Africa, Arabian Peninsula, Madagascar, and the West Indian Ocean islands.
   - 2.3. Oriental region: India, Himalaya, Burma, Malaysia, Indonesia, the Philippines, and the Pacific Islands.
   - 2.4. Australotropical region: northwestern Australia.

3. **Austral kingdom:** southern temperate areas of South America, South Africa, Australasia, and Antarctica. Western portion of Gondwanaland.
   - 3.1. Andean region: South America south of 30° S latitude, extending through the Andean highlands, north of this latitude to the Puna and North Andean Paramo.
   - 3.2. Antarctic region: Antarctica.
   - 3.3. Cape or Afrotemperate region.
   - 3.4. Neoguinean region: New Guinea plus New Caledonia.
   - 3.5. Australotemperate region: southeastern Australia.
   - 3.6. Neozelandic region: New Zealand.

Our goal is to extract the information on area relationships contained in phylogenetic analyses of taxa that live in whole or in part in the Pacific and then summarize them in an area classification. Area relationships or homologs will be tested by area homologs of a range of taxa. We will refine definitions of areas of endemism, as we anticipate that some areas are too large and will need to be subdivided, whereas others are too small and should be combined. Ultimately, the names and definitions of these areas can be standardized under the International Code of Area Nomenclature (ICAN) (Ebach et al., 2008). The Indo-West Pacific, for example, is a broadly recognized name for a distinct biogeographic region; to refer to it by any other name would only diminish our understanding of global biogeography.

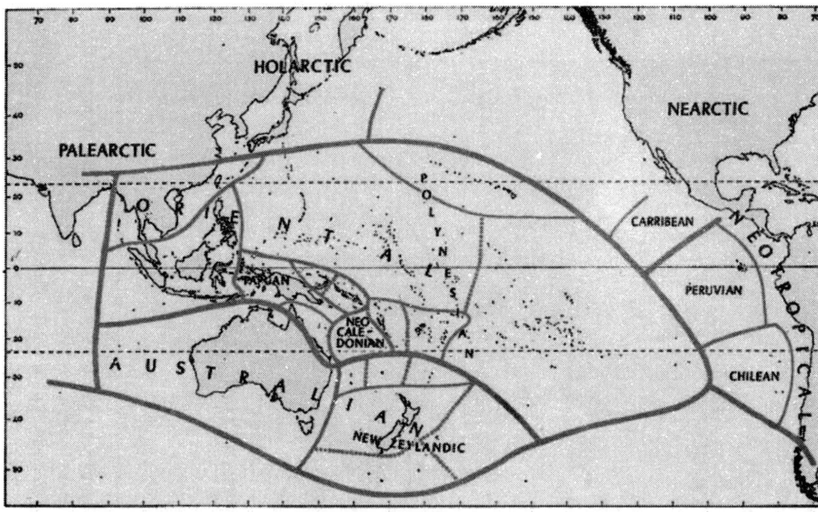

Figure 9.3. Biogeographic areas of the Indo-Pacific based on plant distributions (van Balgooy, 1971: Figure 13). [Modified map reproduced with permission of Peter Hovenkamp.]

Starting Proposal of Areas of Endemism

We begin with a proposal of areas of endemism. Several have been made. The Indo-West Pacific was divided by Ekman (1953) into Indo-Malaya; islands of the central Pacific (except Hawaii); Hawaii; subtropical Japan; tropical/subtropical Australia; and the Indian Ocean. Botanist M. M. J. van Balgooy (1971) catalogued the distributions of Pacific plants and proposed an area summary (Figure 9.3). Despite these proposals of relatively broad regions, it was well understood that finer-scale areas of endemism would be necessary to effectively describe the distribution of taxa, both marine and terrestrial. Finer regions are necessary to propose and test area homologs.

An explicit delimitation of global areas based on the distribution of species of cowries was proposed by Schilder and Schilder (1938–1939; Figure 3.5; see Chapter 3). Ignored by Ekman (1953), the Schilder and Schilder's hypothesis for the Indo-West Pacific was resurrected by Powell (1957; Figure 9.4), who invited discussion of the scheme. Is it useful? Are there too many areas? Too few areas? Are the areas natural? To address these questions, we ask, "What is the relationship among the areas?"

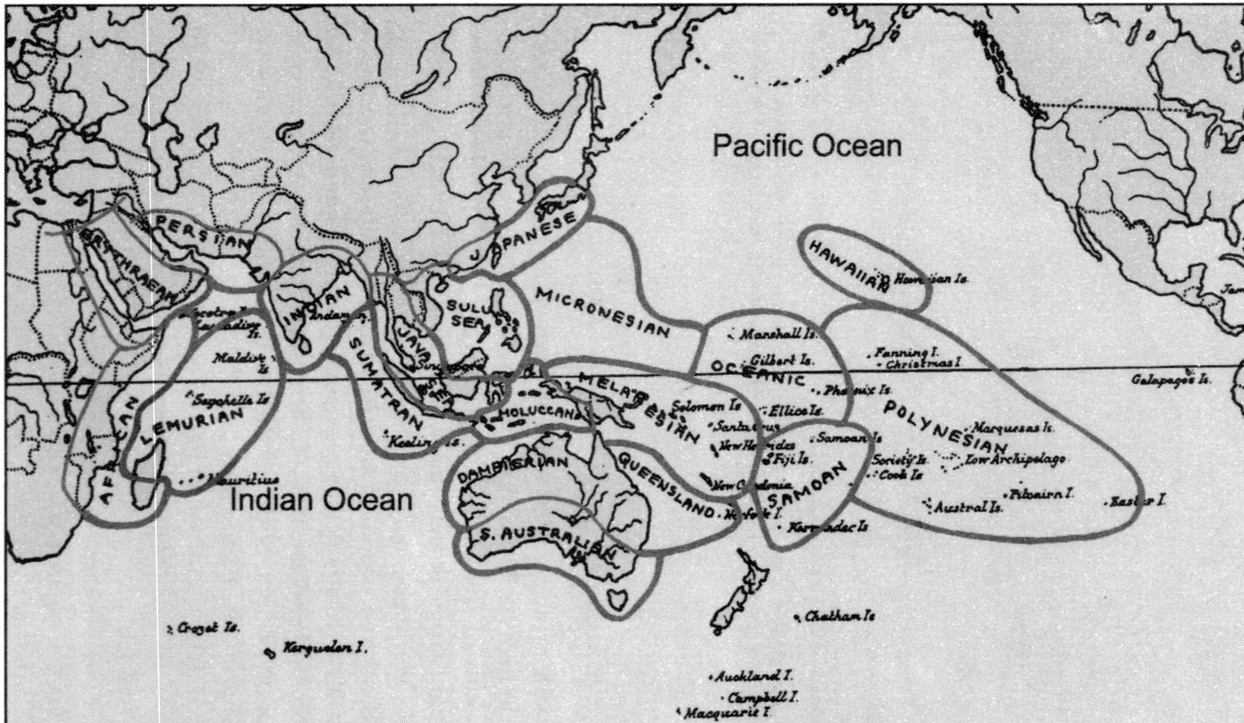

Figure 9.4. Schilder and Schilder's (1938–1939:223: Map 1, in part) 19 areas of endemism throughout the Indo-West Pacific (modified map as redrawn by Powell, 1957). Areas grouped by province are (A) Indian Province: African, Erythraean, Persian, Lemurian, Indian; (B) Central Indo-Pacific Province: Sumatran, Java, Sulu Sea, Japanese, Moluccan, Dampierian; and (C) Pacific Province: Queensland, Melanesian, Micronesian, Oceanic, Samoan, Hawaiian, Polynesian. The 19th region, South Australian, is not included in a province. See text for more details on areas.

TABLE 9.2 THE INDO-WEST PACIFIC REGIONS OF SCHILDER AND SCHILDER (1938–1939) FOLLOWING POWELL (1957), INCLUDING TWO-LETTER AREA CODES IN PARENTHESES

A. Indian Province

1. Erythraean region: Red Sea. **(RS)**
2. Persian region: Persian Gulf to Karachi. **(PR)**
3. African region: Somaliland to Mozambique and southern Madagascar. **(AF)**
4. Lemurian region: northern Madagascar, Réunion, Mauritius, Seychelles, and Maldive islands. **(LE)**
5. Indian region: India and Sri Lanka. **(IN)**

B. Central Indo-Pacific Province

6. Sumatran region: Andaman and Nicobar islands, Sumatra, Christmas Island, Sunda Strait, and south coast of Java. **(SR)**
7. Moluccan region: Bali Strait to Timor, Aru Islands, western New Guinea, and Moluccas. **(MO)**
8. Java Sea region: southern Sulawesi, southeastern Borneo, northern Java, Malay Peninsula, and Gulf of Thailand. **(JS)**
9. Sulu Sea region: Viet Nam, northern Borneo, northern Sulawesi, and Philippines. **(SU)**
10. Japanese region: southeastern China, Taiwan, and southern Japan. **(JR)**
11. Dampierian region: northwestern Australia and Western Australia north of Sharks Bay. **(DA)**

C. Pacific Province

12. Queensland region: New South Wales to Port Curtis Queensland, Lord Howe Island, and Norfolk Island. **(QU)**
13. Melanesian region: northern and eastern New Guinea, Bismarck Archipelago, Solomon Islands, Torres Straits Islands, Vanuatu, New Caledonia including Loyalty Islands. **(MR)**
14. Samoan region: Kermadec Islands, Fiji, Tonga, Niue, and Samoa. **(SM)**
15. Oceanic region: Kiribati and Tuvalu, Tokelau, and Marshall Islands. **(OR)**
16. Micronesian region: Caroline Islands, Palau Islands, Guam, Marianas and Bonin islands. **(MI)**
17. Polynesian region: Cook Islands, Society Islands, Tuamotus, Marquesas Islands, and Rapa Nui. **(PO)**
18. Hawaiian region: Hawaiian Islands and Midway Island. **(HI)**

Area Nomenclature

Eighteen regions, divided into three provinces, were named in the Indo-West Pacific (Table 9.2, from Schilder and Schilder [1938–1939], after Powell [1957:361]).[4] A 19th region, Southern Australia, was mapped by Powell (1957) to complete Australia and agree with Schilder and Schilder (1938–1939; see Chapter 3).

Not all regional names were coined by Schilder and Schilder (1938–1939). Lemuria, for example, was proposed by Sclater (1864) as a "lost continent" that once linked modern Madagascar and India. It was called Lemuria because fossil "lemurs" linked those disjunct areas. Although current tectonic theory does not directly support such a hypothesis, biotic relationships across the Lemurian region endorse its inclusion in our initial proposal. We assign a two-letter code to each region area in our analysis.

Choosing Areagrams

To evaluate Schilder and Schilder's regions and propose relationships among them, we choose areagrams for taxa with endemics in at least three of the areas. Our first example is an areagram for part of the speciose, broadly distributed flowering plant genus *Cyrtandra,* family Gesneriaceae, which provides information on area relationships throughout the western Pacific from Thailand to Hawaii (Cronk et al., 2005; Figure 9.5). Note that most of the names of areas on this areagram are not the same as our 18 Indo-West Pacific regions. To standardize our analysis, we replace the name of the area with the name of the Schilder and Schilder region (Figure 9.6). Some areas, such as "Fiji," are replaced readily by a region, here "Samoan." Others, such as "Australia," are widespread and cover more than one region; when specifying the distribution of *Cyrtandra* species, we replace "Australia" with "Queensland."

The edited areagram (Figure 9.6) contains potential information on relationships of nine of Schilder and Schilder's regions occupied by species of *Cyrtandra*. The areagram written with our two-letter area codes is as follows:

JS(((JS(SU,JS))((QU,MR)(JS,JS)))(SU(SU(JR(HI(PO,SM,SM,MI(SM(PO,PO))))))) Geographic paralogy—repetition of areas—is abundant on this areagram as on many others. "JS" or "Java Sea" occurs five times. How do we extract information on area relationships to discover area homologs? Some area homologs (three-area relationships) can be uncovered "by eye," such as (QU,MR)JS, or (Queensland, Melanesian)Java Sea, and (PO,SM)MI, or (Polynesian, Samoan)Micronesian. The single *Cyrtandra*

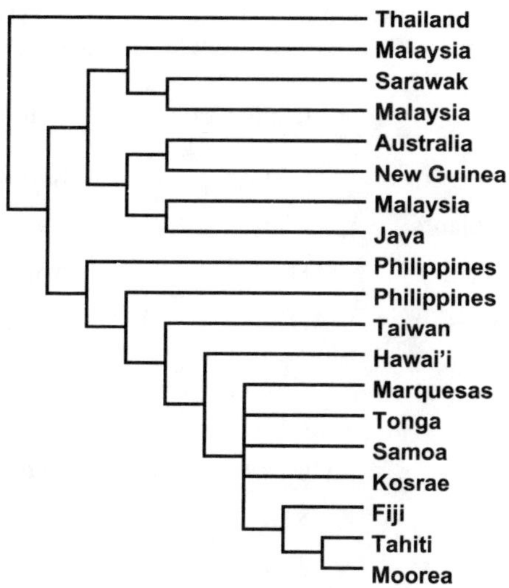

Figure 9.5. Areagram for the genus *Cyrtandra*, following Cronk et al. (2005: Figure 2).

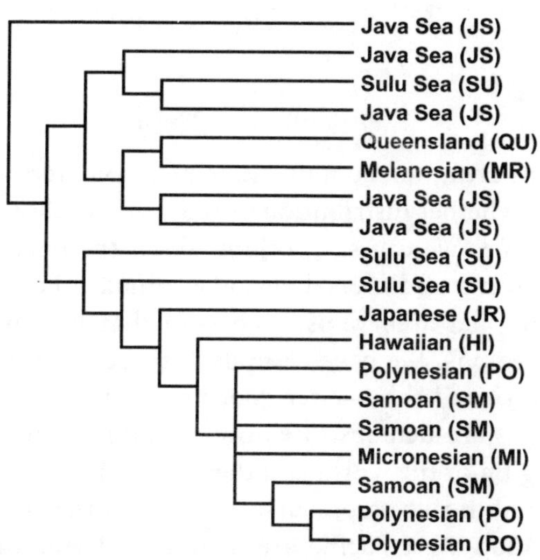

Figure 9.6. Areagram for *Cyrtandra*, modified from Figure 9.3, with area names replaced by the appropriate Indo-West Pacific region of Schilder and Schilder (1938–1939) followed by our two-letter area codes.

areagram thus, at a minimum, implies these relationships for six of the nine areas: ((Queensland, Melanesian), Java Sea), ((Polynesian, Samoan), Micronesian). No explicit area homolog involves the Hawaiian, Japanese, or Sulu Sea regions. None of the areas is widespread on the *Cyrtandra* areagram (Figure 9.6). How we extract area homologs or unique three-area statements from such complex areagrams is covered below.

TABLE 9.3 THREE-LETTER CODES FOR THE 11 CLADES IN THE BIOGEOGRAPHIC ANALYSIS

Taxon	Clade Code	Reference
Cycads	CYC	Keppel et al., 2008
Cyrtandra flowering plants	CYR	Cronk et al., 2005
Halobates water-striders	HAL	Andersen, 1998
Hippocampus seahorses	HIP	Teske et al, 2004
Notograptid fishes	ACA	Mooi and Gill, 2004
Phallostethid fishes	ATH	Parenti, 1989
Pittosporum asterids	COO	Gemmill et al., 2002
Strombus snails	STA	Latiolais et al., 2006
Trimmaton gobies	TRI	Santini and Winterbottom, 2002
Turbinid gastropods	TUB, TUC	Meyer et al., 2005

Geographical Overlap

The *Cyrtandra* areagram contains statements on area relationships but cannot tell us whether distribution of the species follows a general pattern or is unique. We can say nothing about the relative contribution of the mechanisms of dispersal versus vicariance. For this, we need to examine additional areagrams from taxa that live in geographically overlapping regions. We have compiled areagrams from 10 published studies and assigned three-letter codes to clades of Indo-West Pacific taxa (Table 9.3). To demonstrate how to build a biogeographic analysis for one region of the world and then expand it to others, our analysis is conducted in two steps. First, those taxa that live exclusively in the tropical Indo-West Pacific are analyzed. Then, we compare and contrast that result with what we know about area relationships outside of the region.

An areagram for nine species of *Halobates* water-striders was used as an example in Chapter 7 (Figures 7.6, 7.7), following Andersen (1998). The areas in which the water-strider species live need to be translated into Schilder and Schilder's areas to make the water-strider areagram comparable to others. For example, *H. zephyrus* lives in Queensland, and *H. darwini* in the Dampierian region; these were both

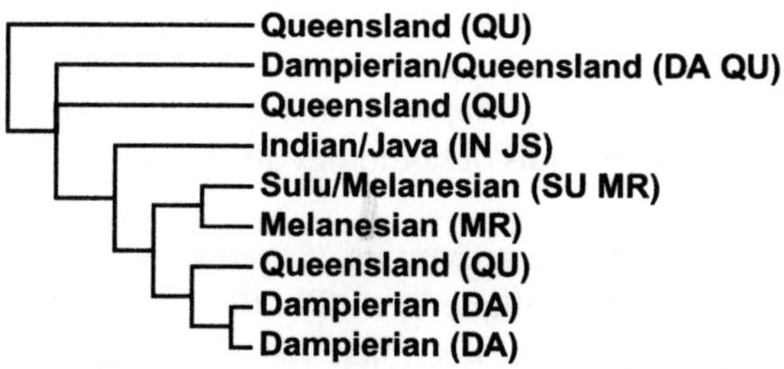

Figure 9.7 Areagram for *Halobates* water-striders, modified from Andersen (1998: Figure 2; Figure 7.7) with area names replaced by the appropriate Indo-West Pacific region of Schilder and Schilder (1938–1939) followed by our two-letter area codes.

called "Australia" by Andersen (1998: Figure 7.7). Andersen (1998) used four areas (Australia, Malayan, Papuasia, and the Philippines) to describe the distribution of *Halobates;* we use six (Queensland, Dampierian, Indian, Java Sea, Sulu, and Melanesian). The edited water-strider areagram (Figure 9.7) demonstrates why area definition is critical, especially when we compare relationships among different taxa that live in the *same* areas: *Halobates* lives in "Australia," specifically Queensland and Dampierian; *Cyrtandra* also lives in "Australia," but just Queensland.

Temporal Overlap

All the taxa in our analyses are recent. We do not need to know the estimated ages of the taxa, if available, before searching for area homologs. We can interpret the sequence and relative timing of events in our general areagram or pattern once it is uncovered. To segregate taxa by estimated age *a priori*—through fossils or a molecular clock, for example—could conceal potentially informative area relationships.

Temporal overlap in biogeography may be part of biotic homology or analogy. Biotic analogy is demonstrated by a biota that shares or has shared the same geographic space with another biota but has a history separate from that biota. Congruence among biotic areas can be tested with general areagrams and with scrutiny of area definition and delimitation, as well as with methods used to estimate ages of taxa. We discuss overlap below.

AREAGRAM ANALYSIS

Each of our clades contains data on area relationships from which we will extract informative statements. We could derive this information by hand, as in some examples above and in previous chapters, but that is time consuming and error prone. Here, we use *Nelson05* (Ducasse et al., 2007), introduced in Chapter 7; this computer program was written expressly for the analysis of homology statements in systematics and biogeography.

To describe how to code areagrams, we return to *Cyrtandra*. Each of the 19 species is given a unique identifier: the letter code for the clade and a number for each of the taxa, here species. Description of the *Cyrtandra* areagram in the *Nelson05* input file is as follows:

CYR1(((CYR2(CYR3,CYR4))((CYR5,CYR6)(CYR7,CYR8)))
(CYR9(CYR10(CYR11(CYR12(CYR13,CYR14,CYR15,CYR16
(CYR17(CYR18,CYR19)))))))

Each clade is described in a similar way: the name of the taxon is replaced with a clade code and a species number.

Distributions are coded by listing the taxa that live in each area. The distribution of *Cyrtandra* species is described for *Nelson05* as follows:

JS:CYR1 CYR2 CYR 4 CYR7 CYR8

SU:CYR3 CYR9 CYR10

JR:CYR11

QU:CYR5

MR:CYR6

SM:CYR14 CYR15 CYR17

MI:CYR16

PO:CYR13 CYR18 CYR19

HI:CYR12

Note that even though geographically paralogous areas, such as "PO" or "Polynesian," are redundant, we do not exclude any of that distributional information. Areas, the taxa that inhabit them, and the relationships among the taxa are input for all clades in the analysis (Table 9.4).

GENERAL PATTERN

General patterns are minimal trees: the most informative summary of area relationships specified by the input areagrams. *Nelson05* uses three-item and compatibility analyses to find the minimal tree, a combination of all

TABLE 9.4. AREAGRAM INPUT FORMAT FOR *NELSON05*: DISTRIBUTION OF TAXA AND RELATIONSHIPS AMONG TAXA FOR 11 CLADES

Areas

AF: TRI1 HIP1 STA1
LE: TRI1 TRI3 TRI4 TRI5 CYC6 ACA3
IN: HAL4 ACA3
SU: CYR3 CYR9 CYR10 HAL5 TRI1 TRI3 TRI5 HIP2 CYC5 ACA1 ACA2 ACA4 ATH2 TUB6
SR: TUB5 CYC4 TUB12
JS: CYR1 CYR2 CYR4 CYR7 CYR8 HAL4 TRI1 TRI3 TRI5 HIP2 HIP3 CYC4 ATH1 TUB4 TUB15
JR: CYR11 HIP3 HIP4 HIP5 STA2 ACA1 TUB10 TUC4 TUB13
MI: CYR16 STA3 STA5 STA7 STA8 STA12 STA13 CYC3 TUB1 TUC2
MR: CYR6 HAL5 HAL6 TRI1 TRI2 TRI5 STA4 STA10 CYC1 CYC2 CYC4 CYC7 COO3 ATH3 TUB3 TUB5 TUB8 TUC5
MO: HIP2 CYC4 ACA2 ACA4 ATH3
SM: CYR14 CYR15 CYR17 CYC2 ACA4 COO1 TUB9 TUC1
OR: STA6 STA9 ACA4 TUB2 TUB7
HI: CYR12 COO2
PO: CYR13 CYR18 CYR19 STA1 TUB11 TUC3 TUB14
QU: CYR5 HAL1 HAL2 HAL3 HAL7 ATH3
DA: HAL2 HAL8 HAL9 TRI1 TRI5

Descriptions

1: (HAL1,(HAL2,(HAL3,(HAL4,((HAL5,HAL6),(HAL7,(HAL8,HAL9)))))))
2: (CYR1,(((CYR2,(CYR3,CYR4)),((CYR5,CYR6),(CYR7,CYR8))),(CYR9,(CYR10,(CYR11,(CYR12,(CYR13,CYR14,CYR15,CYR16,(CYR17,(CYR18,CYR19)))))))))
3: ((TRI1,TRI2),(TRI3,(TRI4,TRI5)))
4: (TUB15,((TUB14,(TUB13,TUB12)),(TUB11,(TUB10,((TUB9,(TUB8,(TUB7,TUB6))),((TUB5,(TUB4,TUB3)),((TUB2,TUB1)))))))))
5: (((TUC1,TUC2),TUC3),(TUC4,TUC5))
6: (HIP1,(HIP2,(HIP3,(HIP4,HIP5))))
7: ((((((STA1,STA2),(STA3,STA4)),((STA5,STA6),STA7)),((STA8,STA9),STA10)),(STA11,STA12)),STA13)
8: ((((CYC1,CYC2),CYC3),CYC4),((CYC5,CYC6),CYC7))
9: (((ACA1,ACA2),ACA3),ACA4)
10: ((COO1,COO2),COO3)
11: ((ATH1,ATH2),ATH3)

NOTE: See Table 9.2 for two-letter area codes in bold and Table 9.3 for three-letter clade codes.

subtrees. A minimal tree is typically more informative than a strict consensus (see also Chapter 7). A strict consensus of two hypothetical subtrees, E(A(BC)) and A(B(CE)), recovers a polytomy or no relationship: ABCE. A minimal tree recognizes the relationship A(BC) shared by both hypothetical subtrees. Area E is uninformative within a minimal tree because it is equally related to all other areas and provides no information on relationship. The same is true for uninformative subtrees (see Box 9.2).

The input areagrams for 11 clades living in 16 areas specified 142 subtrees (Table 9.5). Each of the 11 clades contained one or more subtrees. *Halobates* specified eight subtrees:

(IN,((SU,MR),(QU,DA)))
(IN,(MR,(QU,DA)))
(((SU,MR),(QU,DA)),JS)
(JS,(MR,(QU,DA)))
(IN,((SU,MR),(QU,DA)))
(IN,(MR,(QU,DA)))
(((SU,MR),(QU,DA)),JS)
(JS,(MR,(QU,DA)))

The resulting minimal tree (Figure 9.8) includes all the subtrees that have been reduced to the smallest unit of relationship—a three-area statement. The 142 subtrees included 180 three-area statements. Unlike cladograms of organisms, minimal trees have no branch supports, as the components are based on relationship rather than synapomorphy (= nodes with character states).

The general pattern (Figure 9.8) is well, although not completely, resolved. The 16 areas are related in nine components. Polynesian is excluded from the component that includes the rest of the areas. The component that includes the remaining 15 areas is divided into two broad, overlapping subcomponents: A includes African, Melanesian, Samoan, Micronesian, Oceanic, Hawaiian, Queensland, and Dampierian, and B includes Lemurian, Indian, Sulu Sea, Japanese, Moluccan, Sumatran, and Java Sea.

Polynesian should not be interpreted as primitive or basal to the rest of the areas or as the center of origin for the taxa that live throughout the Indo-West Pacific. This is a general areagram, not a tree. Polynesian is related as in the area homolog: Polynesian(Component A, Component B) (Figure 9.7). Even if one supported dispersal as the only distributional

Box 9.2 Data Dependent, Independent, and Informative

Minimal trees are summaries of informative three-item relationships. The subtree Indian (Micronesian (Queensland, Dampierian)) (Table 9.5, Subtree 2) comprises three independent, informative statements of relationship: Indian (Queensland, Dampierian), Micronesian (Queensland, Dampierian), and Indian (Micronesian, Queensland). These three independent statements added together recover the subtree.

The same subtree, Indian (Micronesian (Queensland, Dampierian)), also includes a single dependent, informative statement, Indian (Micronesian, Dampierian), which is logically implied by the three independent statements.

Dependent statements rely on one or more independent statements to recover a subtree. Here, the area relationships expressed as Indian (Micronesian, Dampierian) are valid, but not independent. The relationship does not conflict with the larger subtree Indian (Micronesian (Queensland, Dampierian)), and does not add any information.

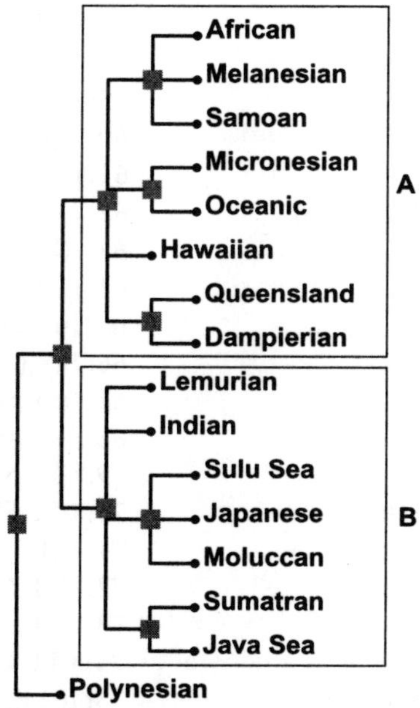

Figure 9.8. Minimal tree, or general areagram, representing the combined area relationships of 11 clades. "Polynesian" should *not* be interpreted as a center of origin (see text for further discussion). Cladogram style as output by *Nelson05*.

TABLE 9.5 142 SUBTREES GENERATED BY NELSON05 FOR THE AREAGRAMS OF 11 INDO-WEST PACIFIC CLADES OCCUPYING 16 AREAS

1: (IN,((SU,MR),(QU,DA)))
2: (IN,(MR,(QU,DA)))
3: (((SU,MR),(QU,DA)),JS)
4: (JS,(MR,(QU,DA)))
5: (IN,((SU,MR),(QU,DA)))
6: (IN,(MR,(QU,DA)))
7: (((SU,MR),(QU,DA)),JS)
8: (JS,(MR,(QU,DA)))
9: (SU,(JR,((MI,(SM,PO)),HI)))
10: (JS,(MR,QU))
11: ((AF,MR),LE)
12: ((AF,MR),(LE,SU))
13: ((AF,MR),(LE,JS))
14: ((AF,MR),(LE,DA))
15: ((AF,MR),(LE,SU))
16: ((AF,MR),(LE,SU))
17: ((AF,MR),((LE,JS),SU))
18: ((LE,MR),SU)
19: ((AF,MR),((LE,DA),SU))
20: ((AF,MR),(LE,JS))
21: ((AF,MR),((LE,SU),JS))
22: ((AF,MR),(LE,JS))
23: ((LE,MR),JS)
24: ((AF,MR),((LE,DA),JS))
25: ((LE,JS),SU)
26: ((LE,MR),SU)
27: ((LE,DA),SU)
28: ((LE,SU),JS)
29: ((LE,MR),JS)
30: ((LE,DA),JS)
31: (LE,(SU,MR))
32: ((LE,JS),(SU,MR))
33: ((LE,DA),(SU,MR))
34: ((LE,JS),SU)
35: ((LE,MR),SU)
36: ((LE,DA),SU)
37: ((LE,JS),(SU,MR))
38: ((LE,SU),JS)
39: ((LE,JS),(SU,MR))
40: ((LE,MR),JS)
41: (((LE,DA),JS),(SU,MR))
42: (LE,(JS,MR))
43: ((LE,SU),(JS,MR))
44: ((LE,DA),(JS,MR))
45: ((LE,SU),(JS,MR))
46: ((LE,SU),(JS,MR))
47: ((LE,JS),SU)
48: ((LE,MR),SU)
49: (((LE,DA),SU),(JS,MR))
50: ((LE,SU),JS)
51: ((LE,MR),JS)
52: ((LE,DA),JS)
53: ((LE,SU),MR)
54: ((LE,JS),MR)
55: ((LE,DA),MR)
56: ((LE,SU),MR)
57: ((LE,SU),MR)
58: (((LE,JS),SU),MR)
59: ((LE,MR),SU)
60: (((LE,DA),SU),MR)
61: ((LE,JS),MR)
62: (((LE,SU),JS),MR)
63: ((LE,JS),MR)
64: ((LE,MR),JS)
65: (((LE,DA),JS),MR)
66: (LE,(MR,DA))
67: ((LE,SU),(MR,DA))
68: ((LE,JS),(MR,DA))
69: ((LE,SU),(MR,DA))
70: ((LE,SU),(MR,DA))
71: (((LE,JS),SU),(MR,DA))
72: ((LE,MR),SU)
73: ((LE,DA),SU)
74: ((LE,JS),(MR,DA))
75: (((LE,SU),JS),(MR,DA))
76: ((LE,JS),(MR,DA))

(continued)

TABLE 9.5 *(continued)*

77: ((LE,MR),JS)	110: ((LE,SU),SM)
78: ((LE,DA),JS)	111: ((LE,SU),OR)
79: ((SR,JR),PO)	112: ((IN,SU),MO)
80: (((((SU,OR),MR),SM),JR),PO)	113: ((IN,SU),SM)
81: ((((SR,(JS,MR)),(MI,OR)),JR),PO)	114: ((IN,SU),OR)
82: ((SR,JR),PO)	115: (LE,(SU,MO))
83: (((((SU,OR),MR),SM),JR),PO)	116: (LE,(SU,MO))
84: ((((JS,MR),(MI,OR)),JR),PO)	117: ((LE,(SU,MO)),SM)
85: ((JR,MR),((MI,SM),PO))	118: ((LE,(SU,MO)),OR)
86: (AF,(SU,(JS,JR)))	119: (IN,(SU,MO))
87: (AF,(SU,JR))	120: (IN,(SU,MO))
88: (AF,(JS,JR))	121: ((IN,(SU,MO)),SM)
89: (AF,(JS,JR))	122: ((IN,(SU,MO)),OR)
90: (AF,((JS,JR),MO))	123: (LE,(SU,JR))
91: (AF,(JR,MO))	124: ((LE,(SU,JR)),MO)
92: ((JR,PO),(MI,MR))	125: ((LE,(SU,JR)),SM)
93: ((MI,OR),MR)	126: ((LE,(SU,JR)),OR)
94: (SR,(MI,MR))	127: (IN,(SU,JR))
95: ((LE,SU),MR)	128: ((IN,(SU,JR)),MO)
96: (JS,(MI,MR))	129: ((IN,(SU,JR)),SM)
97: ((LE,SU),MR)	130: ((IN,(SU,JR)),OR)
98: ((LE,SU),MR)	131: ((LE,(JR,MO)),SU)
99: ((MI,MR),MO)	132: (LE,(JR,MO))
100: ((LE,SU),MR)	133: ((LE,(JR,MO)),SM)
101: (SR,(MI,(MR,SM)))	134: ((LE,(JR,MO)),OR)
102: ((LE,SU),MR)	135: ((IN,(JR,MO)),SU)
103: (JS,(MI,(MR,SM)))	136: (IN,(JR,MO))
104: ((LE,SU),MR)	137: ((IN,(JR,MO)),SM)
105: (MI,(MR,SM))	138: ((IN,(JR,MO)),OR)
106: ((LE,SU),MR)	139: (MR,(SM,HI))
107: ((MI,(MR,SM)),MO)	140: ((SU,JS),MR)
108: ((LE,SU),MR)	141: ((SU,JS),MO)
109: ((LE,SU),MO)	142: ((SU,JS),QU)

mechanism, it would not be possible to say whether taxa dispersed from Polynesian to Component A and B, or the reverse. Also, our choice of areagrams limited the analysis to these 16 areas. Adding more areas, such as those in the eastern Pacific, could provide information that resolves differently the relationships of the Polynesian region.

RECONSIDERED ENDEMIC AREAS

Areas Too Large

Some areas are too large. Recognition of smaller areas could provide additional area homologs. Species in the fish clade ATH live in the northern Philippines (northern portion of Sulu Sea region) and in Palawan (southern portion of Sulu Sea region). Dividing this area into Northern Sulu Sea and Southern Sulu Sea could more finely resolve area relationships.

Areas Too Small

Other areas are too small. The Micronesian and Oceanic regions are sister areas, and thus could be combined into one. They should not be combined with the Polynesian region, as in van Balgooy's scheme (Figure 9.3) because Polynesian is not closely related to the Micronesian and Oceanic regions.

Disjunct Areas

The component containing the African, Melanesian, and Samoan regions is disjunct, spanning the Indian Ocean. Resolving relationships among these three areas would not eliminate the disjunction, but adding taxa that live in intervening areas to the analysis would test the generality of the observation.

Composite Areas

The large islands of Borneo, Sulawesi, and New Guinea, all geologic composites, are biotic composites as well (e.g., Parenti, 1991). Borneo and Sulawesi are part of the Sulu Sea and Java Sea regions, both in Component B, but they are not recovered as sister regions in our analysis. New Guinea is even more disrupted: it is part of the Melanesian region (Component A) *and* the Moluccan region (Component B).

Areas of Overlap

Components A and B overlap broadly (Figure 9.8). Component B is more western, what we could call Asian or Tethyan. But it includes part of New Guinea, a Western Pacific area that has been considered part of the Australian realm. Component A spans the area from the western Indian Ocean to Hawaii, overlapping Component B.

BIOGEOGRAPHY OF THE PACIFIC: PATTERNS AND PREDICTIONS

The complex relationships among the marine and terrestrial biota of the Indo-West Pacific has been summarized in many ways, including by drawing lines, such as Wallace's Line and others between the islands to derive a western or Asian biota from an eastern or Australian, or by recognizing the area between Wallace's and Weber's lines as a separate transition zone named Wallacea (Dickerson et al., 1928; see Parenti, 1991).

The triangle between the Philippine island of Luzon and the Indonesian islands of Sumatra and Timor was called a "hinge between two worlds of life . . . one essentially in tune with the Western Pacific, the other quite as essentially calling into play the quadrangle: Java/Borneo–Japan–Scandinavia–Tanganyika/Natal [the Tethyan region]" (Croizat, 1964:1196). Within the "hinge," life from the two worlds (Asian and Australian or Tethyan and West Pacific, depending on whether one has a terrestrial or marine bias, respectively) either follows tectonic boundaries or overlaps. Alfred Russel Wallace's observation that the Indonesian islands of Bali and Lombok were characterized by Asian and Australian fauna, respectively, is explained as an area of contact between the Eurasian and Australian tectonic plates (see Chapter 2). Overlap is demonstrated by the sympatry of Asian and Australian biotic elements, as for *Acropora* coral species, above. Its explanation is less clear, as it could represent dispersal of elements from one biotic region into another following the contact of tectonic plates or simply repeated distributions that have another explanation. How should this overlap be represented in an area classification?

Schilder and Schilder's three provinces, Indian, Central Indo-Pacific, and Pacific (Table 9.2), are all rejected by the area classification of Figure 9.8. The three provinces were proposed as geographically circumscribed, disjunct areas. Comparing the provincial classification with the area classification highlights the areas of overlap: Component A has representatives in all three provinces, whereas Component B has representatives in the Indian and Central Indo-Pacific provinces, the two major areas of component overlap.

Above, we identified an area homolog: Polynesian(Component A, Component B). This contradicts Ekman's classification of the Indo-West Pacific into the disjunct regions of Indo-Malaya, central Pacific islands (except Hawaii), Hawaii, subtropical Japan, tropical/subtropical Australia, and the Indian Ocean.

Our pattern is a preliminary proposal of area relationships. It needs to be tested with additional taxa and also expanded to include other regions of the Pacific Basin and other regions of the globe. We predict that both marine and terrestrial biota will be equally informative in resolving area relationships.

Our search for a pattern depends on meaningful area delimitation and definition. The practice of using convenient and familiar names such as "New Guinea," "Sulawesi," or "Borneo" for geologically and biologically complex areas of endemism must end if we are to realize our goal of well-corroborated patterns of area relationships. It is alright to use them descriptively, but not as units in a biogeographic analysis. The parallels between systematic biogeography and phylogenetic systematics are deliberate: area homology, like character homology, must be proposed and tested. If necessary, areas, like taxa, need to be redefined and their relationships reconsidered. Classifications of areas, like classifications of taxa, communicate our knowledge of relationships.

SUMMARY

- Complex relationships among the marine and terrestrial biota of the Indo-West Pacific include information about area homology.
- An area classification of the region includes repeated, overlapping patterns that do not agree with disjunct geological areas.
- Systematic biogeography is like phylogenetic systematics in that proposals of area relationship need to be tested and accepted or rejected to form robust area classifications.
- One classification of global regions is (Polynesian(Component A, Component B))(Atlantic, East Pacific))(Boreal, Austral).

NOTES

1. The tropical species classified by Linnaeus (see Chapter 2) were from Indonesia, the Indian Ocean, and the Atlantic and Caribbean, not from the Pacific (see Kay, 1980).

2. Investigations of distribution patterns with the goal of pinpointing the center of taxic diversity have recently focused on the Philippines as the "center of the center of marine shorefish diversity" with geohistorical processes proposed as the cause (e.g., Carpenter and Springer, 2005).

3. The second Mesozoic Tethys Sea is hypothesized to have been formed when the ancient supercontinent Pangea split into Laurasia and Gondwana.

4. The areas of Schilder and Schilder (1938–1939) have been modified by Foin (1976), who considered their relationship to plate tectonics, and by Williams and Reid (2004), for example, who recognized a large, Central Indo-West Pacific region. We begin with the Schilder's smaller areas, as they potentially can lead to a more refined hypothesis of area relationships.

FURTHER READING

Gressitt, J. L. (ed.). 1963. *Pacific Basin biogeography: A symposium.* Bishop Museum Press, Honolulu, Hawaii.

Keast, A., and S. E. Miller (eds.). 1996. *The origin and evolution of Pacific Island biotas, New Guinea to Eastern Polynesia: Patterns and processes.* SPB Academic Publishing, Amsterdam.

Miller, C. B. (ed.). 1974. *The biology of the oceanic Pacific.* Proceedings of the Thirty-third Annual Biology Colloquium. Oregon State University Press, Corvallis.

Trewick, S. A., and R. H. Cowie (eds.). 2008. Evolution on Pacific islands: Darwin's legacy. *Philosophical Transactions of the Royal Society B, Biological Sciences,* 363(1508), 3287–3465.

TEN

THE FUTURE OF BIOGEOGRAPHY

BIOGEOGRAPHY AND IDENTITY

Biogeography incorporates a range of scientific disciplines, including taxonomy, systematics, paleontology, geology, ecology, and evolutionary biology. Because of this naturally broad reach, we are optimistic about the future of biogeography and its prospects for strengthening its position as a truly integrative science.

Biogeography is practiced by a range of scientists who identify themselves as taxonomists, systematists, paleontologists, and so on. Few identify themselves principally as biogeographers, as noted by Nelson (1978; see also Ferris, 1980). This is one reason biogeography suffers from an "identity crisis" (Riddle, 2005). Another reason is the dizzying array of conflicting methods and goals (Chapter 6): "All methods currently employed in cladistic biogeography usually give contrasting results and are theoretically disputed" (Fattorini, 2008:11).

Overview

Biogeographers today are a loosely defined group of researchers without a common aim and language. Biogeography incorporates a range of scientific disciplines, including taxonomy, systematics, paleontology, ecology, and evolutionary biology. For biogeography to become the Big Science that we know it can, biogeographers need to investigate big biogeographical questions (identify common aims) and organize into biogeographical working groups (use a common language).

Working groups, comprising biogeographers from a broad array of taxonomic and scientific disciplines, can tackle challenging biogeographic questions at both local and global scales. Using a single taxon to investigate the distributional history of a set of areas is analogous to using one character state tree to hypothesize the phylogenetic history of a clade: it is incomplete and can mislead. Using a biota to investigate the distributional history of the same set of areas is analogous to undertaking a robust phylogenetic analysis and can lead to robust biogeographic hypotheses.

Biogeographers have proposed a code of nomenclature to aid communication and resolve longstanding conflicts of area names and definitions. Implementation of such a nomenclature beyond small groups of researchers will change the way biogeographers choose and name areas.

Biogeography recognized as an independent science will be able to attract support and be recognized as a field that integrates the biological, geographical, and geological sciences.

This tarnished image of biogeography is ironic. What we call biogeography, in existence since the time of A. P. de Candolle (the elder; Chapter 2), has shaken the foundations of biology and geology by providing crucial evidence in support of evolution and continental drift. It could be so powerful again.

Biogeography can polish its tarnished image. One way is to apply standards. When most people read about biogeography, do a biogeographical analysis, or contemplate which biogeographical method or approach to use, they rarely consider what biogeographical standards have been applied to ensure that they have delimited areas in meaningful ways or chosen an appropriate method. Many methods are not relevant to a comparative biogeography: methods may be either solely ecological or, if comparative, adopted from phylogenetic systematics with little modification (see Chapter 6). This lack of standards is unusual for a comparative science, especially one that depends on empirical data from other fields such as geology and paleontology.

A second way is to be rigorous in framing a biogeographic question and choosing the methods to address it. Biogeographic hypotheses must be tested. Many hypotheses are presented as fact, even called *evidence,* but are not subject to test. One such hypothesis is the Great American Biotic Interchange, which we challenged in Chapter 7.

Distributions are dynamic, they change over time, and our understanding of them may be based on hypothetical models or arbitrary boundaries. Distributions *per se* cannot be studied empirically—only the taxa or biotas that form distributions through time. Using just one taxon to investigate the distributional history of a set of areas is analogous to using one character state tree to hypothesize the phylogenetic history of a clade: it is incomplete and can positively mislead. Using a biota to investigate the distributional history of the same set of areas is analogous to undertaking a robust phylogenetic analysis and can lead to robust biogeographic hypotheses.

Implementing a Comparative Biogeography

Comparative biogeography incorporates systematic and evolutionary biogeography. Systematic biogeography versus evolutionary biogeography reflects an old division between what was called biogeography and chorology, respectively (Haeckel, 1866; see Williams, 2006). The former is the study of distributions, their classification, and their relationships. The latter is the study of individual taxic histories through time and space, forming a synthesis of paleontology, ecology, biogeography, and systematics.

Haeckel's division, added to that between phytogeography and zoogeography, was piled onto a further division of disciplines. Data divide the fields within historical science: if you study rocks, you are a geologist; if you study insects, you are entomologist; and so on. Geologists and entomologists each have departments in which they are separated from botanists who work in herbaria or other departments. Botanists meet at botanical congresses, whereas zoologists split into discipline-specific meetings. The fragmentation of taxonomists and systematists has become more pronounced as the split has widened between those who work exclusively on molecular data and those who do not. All these factors have fragmented systematics to the extent that one group works separately from the other and reads and creates a different literature. The intellectual division has also resulted in more than a little duplication of methods or debate of issues raised in the late 1970s and early 1980s (see Williams, 2006; Parenti, 2006). The difference between areagrams and taxon-area cladograms (TACs) is fundamental in a comparative biogeography, as we have emphasized repeatedly. The difference mirrors that between cladograms and phylogenetic trees, recognized and debated extensively three decades ago in the phylogenetic literature, but dropped from the development and analysis of biogeographic methods.

Methods

Biogeography is commonly classified by its methods (Crisci, 2001; Crisci et al., 2003; Morrone, 2001a). Brooks parsimony analysis (BPA) and parsimony analysis of endemicity (PAE) (see Chapters 5 and 6) both build a matrix of presence or absence data of taxa for a set of areas. Each matrix is analyzed by a parsimony program to find the shortest tree. From a numerical or functional point of view, both methods are similar; BPA differs in that it includes information on nodes. Objectives of the methods also differ. BPA aims to find the mechanism responsible for the evolution and distribution of a particular taxon, whereas PAE groups taxa hierarchically by shared distribution. A classification of biogeographical methods or aims by how they function (e.g., analyzing a matrix with a parsimony program) differs considerably from a classification of methods by their aims or objectives (see Chapter 6).

People have entered biogeography from different fields, and this is reflected in their methods. BPA was developed by a systematic parasitologist investigating co-evolution. PAE was developed by paleontologists with strong backgrounds in taxonomy. Ecological biogeography is largely synonymous with ecology (Nelson, 1978). Phylogeography has the same aims and uses the same methods as phylogenetic biogeography: molecular data are used to infer phylogenetic trees that are in turn used to explain the evolution and distribution of taxa based on Hennig's Progression Rule (see Chapter 6).

Many methods have been tested and compared (e.g., Morrone and Carpenter, 1994; Morrone and Crisci, 1995; Ebach and Edgecombe, 2001; Crisci et al., 2003; Fattorini, 2008; Chapter 6). Biogeography benefits from these critiques. We expect such reviews and analyses to continue, especially as mapping techniques, such as those that incorporate GIS data, become more sophisticated and offer the opportunity to do biogeographic analyses at finer scales.

BIOGEOGRAPHY AS AN INDEPENDENT DISCIPLINE

Biogeography is often erroneously treated as a subdiscipline of evolutionary biology. Yet the study of distribution patterns supports evolution. Biogeography began more than a century before evolutionary biology. Biogeography has literally rocked geology and developed evolutionary theory as only a strongly independent field can.

A Biogeographical Question

Every independent field in historical science asks its own questions. Geologists ask, "What is a rock?" A rock is a collection of minerals. This leads to a classification of rocks and minerals. Each discipline has a nomenclatural code and classification. A geologist discovering a rock formation in the field asks, "What kind of rock is this?" In so doing, geologists rely on the mineralogical and petrological classification system to identify the rock and, most importantly, to date the rock according to the principles underlying stratigraphy and geochemistry. The formation of the rock also has a history. Rocks, like all forms, have distinctive structures resulting from interaction with the environment. Deformation of the rock tells us about the history of the terrane to which it belongs, and that, in turn, tells of the geological history of the area as a whole.

Taxonomists ask, "What is a taxon?" and discover and describe diversity. Taxa are placed in a classification, again following a nomenclatural code. Systematists ask, "What is a clade?" and discover monophyletic groups and evolutionary relationships.

Areas of endemism are the building blocks of biogeography. Without them, biogeography would not exist. A purely biogeographical question, but one that not all biogeographers ask, is *"What is an area?"*

What Is an Area?

In biogeography, an area is the place occupied by a biota: a group of taxa and their combined distribution circumscribed by geographical limits. Abiotic areas belong exclusively to geology and geography. A biotic area may be placed in a classification system (see Chapter 2). The system we use today is built on those of Lamarck and Candolle (1805) and Sclater (1858). Biotic areas have characters that make them distinct from other biotic areas, and, most important, biotic areas share relationships, and therefore, a history.

Relationship among biotic areas has not been considered central to biogeography and its related fields. By acknowledging that biotic areas have relationships and through these relationships can and should be classified, we recognize a field that can ask big questions.

BIOGEOGRAPHY AS AN INTEGRATIVE DISCIPLINE

Biogeography can be integrated by implementing a standard that can be used by all its practitioners. To ask big questions, we need to organize as biogeographers, not along taxonomic or methodological lines, but according to the *areas* we study. An island biogeographer may have little use for subtree analysis, and a cladistic biogeographer may reject the notion of centers of origin; together, as biogeographers, they may study the same area, say the Caribbean Basin or the Pacific Plate. *Biogeographers are united by the areas that they study.* When we move away from the needless and fragmentary divisions between zoogeographers, phytogeographers, and paleobiogeographers, we can recognize and support groups such as Pacific Basin biogeographers or Tropical Lowland South American/African biogeographers, and so on.

Studying an area, be it a large region such as Australasia or a small one such as a Pacific atoll, has its challenges. Most biogeographers work independently on their taxa and on accumulating distributional data that may be used in a biogeographical analysis. Ornithologists who study the distributional history of their taxa will not necessarily be able to describe the entire biotic area in which their birds live; most such studies do not ask a biogeographical question but provide hypotheses to address phylogenetic, evolutionary, or ecological problems. A single biogeographic dataset has little power alone to generate a robust distributional hypothesis; it can be added to others in a comparative biogeography.

Through organization and application of standards unique to biogeography, its disparate parts can become integrated. Biogeographical working groups may be united or integrated in several ways (see Box 10.1).

Box 10.1 A Guide to Integration

1. Biogeographers can organize into biogeographical working groups to ask questions relevant to a particular biotic area.
2. Big questions can be posed to test and challenge current theories within geology and biology.
3. Standards (e.g., nomenclature) can be established to help biogeographers communicate.
4. Biogeographical theories can be proposed that inform big questions in other fields, such as geology.
5. Integration does not mean assimilation.

Aims

Biotic areas share relationships with other biotic areas. What do these relationships tell us? How can we use these relationships to make hypotheses about the geographical and geological evolution of an abiotic area?

Once we combine our biogeographical data and find a general pattern, we can consider distributional mechanisms such as disjunction, migration, or extinction. A biogeographical working group can be united around these aims. Geologists have working groups within the UNESCO International Geological Congress Project (IGCP). Each project has a goal to be investigated over a certain time period. Geological working group IGCP 509 is a group of "150 geoscientists from more than 25 countries, sharing a common goal of understanding the evolution of our planet through its middle age: the Palaeoproterozoic Era (2500–1600 Ma)."[1] Analogous principles can be applied to form a biogeographical working group.

A Common Language for Biogeography

Even if we all were to apply one method, biogeography would not be unified or integrated. For this, we need a common language. The lack of a standardized naming system or language for areas is one of the biggest problems faced in biogeography. A biogeographical nomenclature is vital to communicate ideas or organize a biogeographical working group.

Language unifies biogeography as it does all fields. A paleontologist may ask a mineralogist, "What is diorite"? The mineralogist will be able to state clearly and concisely that diorite is "medium- to coarse-grained intrusive igneous rock that commonly is composed of about two-thirds plagioclase feldspar and one-third dark-colored minerals, such as hornblende or biotite."[2] Furthermore, the mineralogist will be able to point to where diorite occurs in the Quartz, Alkali feldspar, Plagioclase, Feldspathoid (QAPF) graph, a granite classification system. Geology has a standardized language with which geologists can communicate effectively. We have no such system in biogeography. A macroecologist may tell a panbiogeographer, "This eucalypt lives in Australia." Given that Australia may be "all known areas in the Australian land-mass" (Seberg, 1991) or "all [of those] areas except New Guinea and Queensland" (Wagstaff and Dawson, 2000), neither the panbiogeographer nor the macroecologist will know exactly where in "Australia" that eucalypt lives.

To be meaningful for comparative biogeography, areas need to be proposed, named, and classified based on a set of geographical limits and a list of endemic taxa. An ideal nomenclature for biogeographical classification would require the same rules as those applied in taxonomy (i.e., International Commission on Zoological Nomenclature, or ICZN), in which names require diagnoses and descriptions. An International Code of Area Nomenclature, or ICAN, has been proposed by the Systematic and Evolutionary Biogeographical Association (Ebach et al., 2008; see Chapter 3). With a nomenclature in place, it will be possible not only to communicate effectively, but also to store data more efficiently in electronic repositories such as museum databases. Existing structures (e.g., taxonomic databases) use geographical and "biogeographical" regions to categorize data. Locality data vary from vague to precise. Yet a biogeographical database is in demand, especially for geographical and descriptive tools. Once the principles of an ICAN are implemented, biogeographers will be able to assist other fields reliant upon unambiguous biogeographical names.

Large working groups have recently published comprehensive, standardized classifications of global ecoregions that map biodiversity for a variety of purposes, including generating conservation plans and increasing biogeographic literacy. Three of the largest are Terrestrial Ecoregions of the World (Olson et al., 2001), Marine Ecoregions of the World (MEOW) (Spalding et al., 2007), and Freshwater Ecoregions of the World (FEOW) (Abell et al., 2008). These are monumental efforts that will form the basis of area identification and nomenclature for future biodiversity and conservation efforts. We comment on their role in a comparative biogeography.

Consider the Madeira-Tapajós moist forest of South America of Olson et al. (2001). The Madeira-Tapajós moist forest (NT0135) is in Central Amazonia, in Brazil and parts of Bolivia. This forest is 277,900 square miles (719,700 square kilometers) in area—about the size of California and New Mexico combined—and the biome is classified as "Tropical and Subtropical Moist Broadleaf Forests." The description continues:

> This ecoregion spans a huge, diverse area in central Amazonia, extending across the lowland Amazon Basin south of the Amazon River, and reaching south to the border between Brazil and Bolivia. Most of this region receives between 80 and 120 inches (2,032–3,048 mm) of rainfall each year, however, as much as 160 inches (4,064 mm) of rain annually drenches the middle Madeira River section! And, along the northern and southern edges of the region, annual rainfall is less than 80 inches (2,032 mm). . . .
>
> Many different habitats stretch across this huge area, each with a unique type of vegetation. Dense lowland and premontane rainforest, open-canopy premontane rainforest, woodland savanna, grasslands, and semi-deciduous forest all grow here. White-sand igapó forests grow where blackwater rivers, those that lack a sediment load, flood the region. In these forests, large trees such as piranheira, louros-de-igapo, and jacareúbas pierce the sky. The dense lowland rainforest reaches 100 feet (31 m), with some emergent trees rising 50 more feet (15 m) above the carpet of green crowns. The very rare, locally endemic *Polygonanthus* tree can be found in the lowland rainforest near Maués. In the hill woodlands of upper Marmelos, the dwarf rubber tree, the elegant palm, the enormous *Huberodendron* tree, and the rare *Brachynema* grow (www.nationalgeographic.com/wildworld/profiles/terrestrial/nt/nt0135.html).

Description of the Madeira-Tapajós moist forest (NT0135) is highly detailed and includes statements on endemic taxa as well as habitat variation. Areas of endemism can likely be identified within the moist forest and their relationships investigated. Yet it is a vast area that likely is a composite of many subareas, all of which together may not form a monophyletic area, in a cladistic or comparative biogeographic sense.

Descriptions of areas in MEOW and FEOW are similarly detailed where the data exist. Yet each classification has its drawbacks. FEOW is based solely on freshwater fish distributions. Species composition, not endemism, is key to identifying areas in MEOW. Neither scheme addresses relationships among areas at either local or global scales, although ecoregions in each are grouped into 12 larger regions or realms. A comparison of the 12 global realms and coastal and shelf regions of MEOW to the 12 global regions and coastal regions of FEOW could be

Box 10.2 Organizing a Biogeographical Working Group Project

1. Choose clades that have overlapping distributions (e.g., genera that share similar geographical distributions).
2. Define the study area based on biological as well as geological, geographical, and ecological elements.
3. Investigate the inorganic history of an area based on biotic relationships as discovered through general areagrams.
4. Relate general areagrams to those in other areas of the world (i.e., place the area under study in a regional or global classification).
5. Devise ways to test and modify the classification.

illuminating. Such a comparison among freshwater and marine regions would likely lead to revisions of both schemes and provide the first of many tests of the area classifications.

Biogeographers, unified through a common language of nomenclature, can form biogeographical working groups that target biodiversity and conservation. Large-scale biodiversity projects are often organized around taxa (e.g., floral or faunal studies), rather than around areas. Some organized around areas may not be biogeographically meaningful (e.g., political areas or grids). A biogeographical working group on Melanesia may assess the systematics and ecology of a region rich in biodiversity. A study that involves extensive fieldwork, including ecological assessments, specimen collecting, and mapping, will help systematists and biogeographers collaborate to identify, classify, describe, and revise taxa from biologically diverse global regions (see Box 10.2). The resulting areagrams will reflect the combined effort of ecologists, systematists, and biogeographers, ensuring that each area within the region is adequately defined, described, and assessed.

We are optimistic that global projects can accelerate species discovery, taxon and area descriptions, and cybertaxonomy and can set a standard for collaborative scientific research in the 21st century. Future biogeographers can be well-funded planetary biologists, leading large scale global initiatives, discovering life from the highest mountain peaks to the lowest depths of the oceans and discovering and unraveling the interrelationships of our planet's biota. Area classifications can become the organizing frameworks for biogeographical data at local

Box 10.3. Biogeography as Big Science

The term *Big Science* was coined by Alvin M. Weinberg (1961) for large-scale science efforts, usually those associated with technology (e.g., genomics, genetic engineering, space exploration) or medicine (e.g., AIDS or cancer research). Big Science has since been expanded to include large-scale funding and overarching questions in the historical sciences, such as geology and astronomy.

Biogeography has not been thought of as Big Science, but it can come to be so by organizing around three points:

1. Big questions. Biogeography is able to ask big questions that organize information on distributions in new ways that can challenge our current hypotheses, or it can directly address topics once considered to be outside its scope (i.e., it can redirect questions about geology of the Pacific, Chapter 9).
2. Big working groups. An organized group of biogeographers established at a school, university department, or institute can address questions that one biogeographer working alone could not answer.
3. Big budgets. Long-term national or international consortium funding (e.g., NSF or NIH in the US; EU in Europe; ARC in Australia) can ensure the future of biogeography and biogeographers.

and global scales. These will become understood as classifications not of geological, geographic, or political areas, but of biotic areas that demonstrate the relationship between organisms and the environment through time.

SUMMARY

- The study of biotic areas unites biogeography. Yet biogeographers are divided by their objectives: to describe and empirically classify biotic areas or to explain the distribution mechanisms of individual taxic histories.
- A standard naming system, such as the ICAN, can help unite biogeographers to resolve area delimitation and definitions.

- Biogeography potentially can unite its practitioners by asking big questions, by attracting big budgets, and by establishing big working groups.

NOTES

1. http://earth.geology.yale.edu/igcp509/
2. Definition of diorite from *Encyclopaedia Britannica* online (http://www.britannica.com/eb/article-9030562/diorite#987.hook), accessed 30 April 2008.

FURTHER READING

Lomolino, M.V., and L.R. Heaney. 2004. Reticulations and reintegration of modern biogeography. In M.V. Limolino and L.R. Heaney (eds.), *Frontiers of biogeography: new directions in the geography of nature* (pp.1–3). Sinauer Associates, Inc., Sunderland, Massachusetts.

Price, D. J. D. S. 1963. *Little science, big science.* Columbia University Press, New York.

Glossary

ALLOPATRY (ALLOPATRIC) The **distribution** of two or more **taxa** in disjunct, non-overlapping areas. See also **sympatry (sympatric).**

ANCESTRAL AREA ANALYSIS (AAA) A biogeographic method that assigns a value or a loss to a series of **paralogous nodes** within a **taxon-area cladogram (TAC)**. An unbroken series of paralogous nodes closer to the basal node are given a gain, whereas those in a broken series acquire a loss.

ANTITROPICAL (BIPOLAR) A **distribution** characterized by **taxa** living in the holarctic (boreal) and austral regions of the world, but not in the intervening tropical region.

AREA 1. A *geographic area* is any area delimited by geographic (e.g., shoreline of a lake or ocean; mountain range; river basin) or geopolitical (e.g., state border line) boundaries. 2. A *biotic area* is the geographical area inhabited by a **biota.** Limits of **taxon distribution** specify limits of the area. **Biotic areas** form the basis of **biogeography**.

AREA ANALOGY A statement of **similarity** between two or more areas that share a closer **relationship** with other areas than to each other.

AREA CLADISTICS A method that interprets the **relationships** among **biotic areas** on an **areagram** as proximal geographic distances.

AREA CLADOGRAM See **areagram.**

AREAGRAM Also called an area cladogram; any branching diagram, graph, or written statement that depicts the **relationship** among three or more **areas.**

AREA HOMOLOG A statement of **relationship** among three or more **biotic areas** (i.e., a three-area statement). Two or more area homologs that share the same relationship demonstrate **area homology.**

AREA HOMOLOGY Also known as *geographic congruence*; a discoverable geographic pattern that consists of two or more **area homologs** that share a set of **relationships** (i.e., three-item relationships). Area homology is evidence of **biotic area** evolution (a shared or common history).

AREA OF ENDEMISM An area characterized by the overlapping **distributions** of two or more **taxa.**

AREOGRAPHY See **chorology.**

BARRIER A physical obstacle that is observed or inferred to prevent or impede the movement of organisms.

BATHOLITH A large **igneous** intrusion (molten body of rock) that forms inside the Earth's crust.

BIOCHORE In paleobiogeography, the biogeographic units recognized as **areas of endemism** through time.

BIOGEOGRAPHIC NODE In **panbiogeography,** the point at which two or more **tracks** intersect.

BIOGEOGRAPHY 1. The study of the **distribution** of life. 2. Systematic biogeography is the systematic study of **biotic area relationships** and classification. 3. Evolutionary biogeography is the study of **taxic** distributional mechanisms.

BIOTA A group of **taxa** (organisms), the combined distribution of which occupies a common set of geographical limits.

BIOTIC AREA See **area.**

BROOKS PARSIMONY ANALYSIS (BPA) A matrix-based method that codes the absence or presence of areas on the **nodes** of one or more **taxon-area cladograms (TACs).** The absence/presence in a node-by-area matrix is analyzed using a parsimony program.

CHOROLOGY The study of **taxa** and their individual **distributional** histories. See also **evolutionary biogeography.**

CLADISTIC METHOD Any method that attempts to find the **relationship** between two or more **taxa** based on **homologs.**

CLADISTICS The systematic study of **taxic** classifications and their **relationships.**

CLADOGRAM Any branching diagram, graph, or written statement that depicts the **relationship** between three or more **taxa.**

CLASSIFICATION 1. Artificial: any hierarchical key based on **taxic** similarity used for the purposes of identification. 2. Natural: see **monophyly (2).**

COMPARATIVE BIOGEOGRAPHY The comparative study of **biotic areas** and their **relationships** through time.

COMPARATIVE BIOLOGY The comparative study of organisms (and their parts) and their **relationships** through time.

COMPARATIVE PHYLOGEOGRAPHY See **phylogeography.**

COMPONENT A junction within an **areagram** that specifies what **areas** are present in the included terminal branches.

COMPONENT ANALYSIS A **cladistic** method that uses **components,** rather than **nodes** or data matrices, to resolve **area relationships.**

CONTINENTAL DRIFT A geological process in which present-day continents were once located in different parts of the globe, likely as a single landmass (e.g., **Pangea**), and drifted into their current positions.

COSMOPOLITANISM The hypothesized widespread ancestral **distribution** of a **taxon** prior to **vicariance** and regionalization.

DISPERSAL 1. *Descriptive dispersal or dispersion* states that any given **taxon** or **biota** moves within its *potential* **distributional** range, although no **explanation** or **mechanism** is given. 2. *Explanatory dispersal* states that a particular set of historical events and physiological mechanisms involving the dispersion of a particular taxon or biota (i.e., jump dispersal, long-distance dispersal, and so on) is responsible for genetic isolation and speciation.

DISTRIBUTION The geographical area occupied by a **taxon** or a **biota.** A distribution can be known through historical records and sampling or through estimation of a potential range.

EARTH 1. The third planet from our Sun. 2. Dynamic Earth: a concept which states that Earth is changing rapidly (see **plate tectonics**). 3. Static Earth: a concept which states that Earth is changing slowly.

ECOLOGICAL STRANDING A **mechanism** that results in changes in the habitat of a **biota** over time without requiring movement or **dispersal** of members of the biota.

ECOLOGY The study of the interaction between organisms, or between organisms and the physical environment.

ENDEMIC AREA The geographical **area** to which a **taxon** or **biota** is understood to be native.

ENDEMISM A **taxon** is said to be endemic to an **area** if it lives there and nowhere else.

EPICONTINENTAL SEA A large body of seawater confined to a continental plate.

EROSION The physical action of moving weathered sediment away from its source. See also **weathering.**

EXPANDING EARTH THEORY A theory that attempts to explain **continental drift** by proposing that the Earth, formerly consisting of a single continental landmass, has expanded through the constant creation of oceanic crust. See also **tectonics.**

EXPLANATION A hypothesis or reason given for an observed **process** or discovered **pattern.**

EXTINCTION The demise of an organism, **taxon,** or **biota.**

FORMATION In geology, a unit of **rock.**

FOSSIL (ORGANISM) Remains of an organism that lived in a previous geological period, preserved in various ways, such as mineralization of a skeleton or imprint on **rock.**

GENEALOGY 1. The **relationships** between a parent and its offspring, their extended family, and distant relations that are recorded or observed. 2. The unobserved and unrecorded hypothetical relationships between members of a population over generations.

GENERAL AREAGRAM Any branching diagram, graph, or written statement that depicts geographical congruence or **area homology.**

GENERALIZED TRACK The overlap of one or more **tracks**; also called a **standard track.**

GENERAL LAW A theory that uses an explicit **mechanism** to explain a **process** or **pattern.** Laws contain a list of given *exceptions*, which make them immune to any contrary evidence or **explanation.** See also **universal system.**

GEOCHEMISTRY The study of chemical composition and chemical changes of terrestrial or extraterrestrial minerals or **rocks.**

GEOGRAPHICAL INFORMATION SYSTEM (GIS) Any method or device which captures, stores, or analyses geographical coordinates.

GEOLOGICAL CLADOGRAM Any branching diagram, graph, or written statement that depicts **similarities** or proximal distances among abiotic **areas** in time. See also **terrane analysis.**

GRADE A group of unrelated **taxa** or **biota**. See also **paraphyly**.

GRADISTICS The use of **grades** to classify or define phylogenetic or genealogical lineages.

HISTORICAL AND ECOLOGICAL BIOGEOGRAPHY A classification between biogeographical and ecological biogeographic methods, respectively.

HISTORICAL SCIENCE A non-experimental biological, Earth, or physical science that examines past events or **relationships** that may reveal common patterns.

HOMOLOG The manifestation of the same form (e.g., human forearm, bird wing, dolphin fin) within a group of related organisms (e.g., tetrapods).

HOMOLOGY A **relationship**.

HOMOPLASY A non-homologous grouping based on **similarity**. See also **analogy** and **paraphyly**.

IGNEOUS ROCK Any agglomeration of minerals that were formed from molten material inside (intrusive) or outside (extrusive) the Earth.

INTERNATIONAL CODE OF AREA NOMENCLATURE (ICAN) A naming system for biogeographic **areas**. The code specifies that a named area be linked to a published diagnosis or description.

ISLAND BIOGEOGRAPHY The geographical study of populations and species richness, especially of discrete **areas** such as islands or isolated mountaintops.

MAP 1. *Distributional or geographical maps* depict the known geographical **distribution** ranges of one or more named **taxa**. 2. *Biogeographical maps* depict the regional distribution and classification of **biota**.

MECHANISM 1. Observed: any observed and repeatable (i.e., testable) chemical or physiological reaction that is responsible for a certain **process** or **pattern** (e.g., photosynthesis). 2. Explanatory: any hypothetical (unobserved, untestable) mechanism proposed to explain a process or pattern.

METAMORPHIC ROCK A **rock** that has been deformed through heat and pressure to the degree that it has acquired its own structural and mineralogical characteristics.

MINERALOGY The study of minerals.

MONOPHYLY 1. A group of all known descendants and their most recent common ancestor. 2. A **relationship**.

MORPHOLOGY The study of form based on the behavior, shape, hard or soft parts (e.g., histology, osteology, behavior, carapace), and genetics (e.g., alleles, chromosomes, mitochondrial and ribosomal DNA) of living, **extinct**, or **fossilized** organisms.

MULTIPLE AREAS ON A SINGLE TERMINAL-BRANCH (MAST) Multiple (more than one) **biotic area** on a single terminal branch in an **areagram** or **taxon-area cladogram (TAC)**.

NODE 1. A junction or terminal in a **taxon-area cladogram** (TAC) that represents phylogenetic data (e.g., character state **distribution**). 2. A hypothetical ancestor or historical event (e.g., **extinction**). 3. A synapomorphy on a **cladogram.** 4. An **area** where two or more **generalized tracks** overlap.

PANBIOGEOGRAPHY A comprehensive method of global biogeography first outlined by Léon Croizat (see Chapter 1) characterized by depicting the repeated, overlapping **distribution patterns** or **tracks** of many unrelated organisms on **maps.**

PANGEA The ancient supercontinent that comprised all known global land masses. Under the theory of **continental drift,** Pangea began to break up during the Jurassic period (200–145 mya).

PANTROPICAL A **distribution pattern** characterized by organisms living throughout the tropical region of the world but being absent from the holarctic (boreal) and austral regions.

PARALOGY-FREE SUBTREE ANALYSIS A method that uses **area relationships** to remove **geographic paralogy.**

PARALOGY (GEOGRAPHIC) The duplication of **biotic area** names on an **areagram.** Paralogous **components** are uninformative about area **relationships** and may occur in **MASTs.**

PARAPHYLY, PARAPHYLETIC 1. A group of organisms that does not include all descendants and a common ancestor. 2. An artificial or non-evolutionary group consisting of organisms that are more closely related to other organisms than they are to each other.

PARSIMONY The logical criterion applied in **systematics** and **biogeography** that obtains a **cladogram** from a given set of characters that is the best supported by those characters.

PARSIMONY ANALYSIS OF ENDEMICITY (PAE) A method that codes the absence and presence of **taxa** in a series of **areas** or geographical units (e.g., geopolitical areas or grids). The absence/presence-by-area matrix is analyzed using a **parsimony** program.

PATTERN The recognition of more that one **homolog** supporting a **taxic** or **biotic relationship.** See also **monophyly** and **area monophyly.**

PETROLOGY The study of **rock** classification.

PHYLOGENETIC BIOGEOGRAPHY A method that uses **taxon-area cladograms (TACs)** to reconstruct the history of a particular lineage or **genealogy** using a particular hypothesis (e.g., **Progression Rule**). **Phylogeography** differs in that it largely uses molecular data and is focused at the species or population level.

PHYLOGENETIC TREE A hypothetical **genealogy** of **taxa** usually depicted as a branching diagram.

PHYLOGENY 1. A hypothetical **genealogy** of **taxa.** 2. A set of taxic **relationships** that represents a natural grouping (**monophyletic** group). See also **homology.**

PHYLOGEOGRAPHY See **phylogenetic biogeography.**

PLACE A physical environment in which organisms live or have once lived.

PLATE TECTONIC THEORY A theory that attempts to explain **continental drift** by stating that the Earth is divided into continental and oceanic plates, which rift apart, especially in the sea floor, or are subducted, one under the other, at plate margins.

PROCESS Any observed, recorded, and repeatable series of events that leads to the formation of a **pattern** (e.g., ontogeny).

PROGRESSION RULE A hypothesis central to **phylogenetic biogeography** or phylogeography which states that the most basal **node** on a **taxon-area cladogram** (TAC) is the center of origin (or part of an ancestral **area**) and that subsequently more derived **taxa dispersed** away from that center of origin.

RECIPROCAL ILLUMINATION 1. The same **pattern** discovered by two or more different types of evidence acquired by different means (e.g., geological reconstruction and general congruence). 2. The same explanatory hypotheses originating from two or more different fields (e.g., paleontology and molecular **systematics**).

REDUCED AREA CLADOGRAM A **paralogy**- and **MAST**-free **areagram.**

REFUGIUM An **area** that maintains a relictual population of a previously more widespread **taxon.**

RELATIONSHIP 1. When two characters, **taxa,** or **biota** are discovered to be manifestations of the same form when compared to a third. 2. A narrative based on an unobserved and hypothetical **genealogy** (i.e., ancestor–descendants) or **phylogeny** (e.g., ghost lineages).

ROCK An agglomeration of minerals.

SEDIMENTARY ROCK A **rock** that is formed by sediment usually from other rocks.

SIMILARITY The comparison of any two objects based on a list of qualities or quantities.

SUBTREE ANALYSIS (PARALOGY-FREE) A method that recovers **relationships** within an **areagram** that contains **paralogy** and/or **MASTs.**

SYMPATRY (SYMPATRIC) The **distribution** of two or more **taxa** in the same or overlapping ranges.

SYSTEMATICS The study of classification based on comparing **taxic relationships** or similarities.

TAXON A named group of organisms.

TAXON-AREA CLADOGRAM (TAC) A **phylogenetic tree** in which the names of the **taxa** are replaced with the names of the **biotic areas** in which they live.

TAXONOMY The study of biological naming and classification.

TELEOLOGY Invoking a purpose as an **explanation** of a biological form or natural **process.**

TEMPORAL PARALOGY A technique used to date the **nodes** within a branching diagram based on the age of **taxa** or **biota** found at the nodes.

TERRANE ANALYSIS A biogeographic method that uses **general areagrams** (**area homology**) to choose the most relevant hypothetical **geological cladogram.**

TRACK A line or graph drawn on a **map** that links the **areas** of **distribution** of a **taxon** or group of taxa.

TRANSPARENT METHOD A method that attempts to resolve **MASTs** into one or more possible **relationships** prior to implementing **paralogy-free subtree analysis.**

VICARIANCE 1. *Descriptive vicariance* states that **taxa** or **biota** are unable to freely **disperse** within their potential **distributions** due to a geographical **barrier,** although no particular **mechanism** or event may be specified. 2. *Explanatory vicariance* states that a particular set of historical events and abiotic mechanisms, which geographically divide a taxon or biota, are responsible for genetic isolation and speciation.

VICARIANCE EVENT Any natural formation of geology or geography (e.g., vulcanism, mountain building, earthquake, glaciation, stream capture) that results in the splitting of a **taxon** or **biota** into two or more geographical regions.

WEATHERING The physical or chemical processes involved in the disintegration of **rocks** into sediment (e.g., gravel, pebbles, sand).

Bibliography

Aalbu, R.I., and F.G. Andrews. 1985. New species relationships and notes on the biology of the endogean tentyrine genus *Typhlusechus* (Tenebrionidae: Stenosini). *Occasional Papers in Entomology*, 30, 259–28.

Abel, O. 1912. *Grundzüge der paläobiologie der wirbeltiere*. Schweizerbart, Stuttgart.

Abell, R., M.L. Thieme, C. Revenga, M. Bryer, M. Kottelat, N. Bogutskaya, B. Coad, N. Mandrak, S. Contreras Balderas, W. Bussing, M.L.J. Stiassny, P. Skelton, G.R. Allen, P. Unmack, A. Naseka, R. Ng, N. Sindorf, J. Robertson, E. Armjio, J.V. Higgins, T.J. Heibel, E. Wikramanayake, D. Olson, H.L. López, R.E. Reis, J.G. Lundberg, M.H. Sabaj Pérez, and P. Petry. 2008. Freshwater ecoregions of the world: A new map of biogeographic units for freshwater biodiversity conservation. *BioScience*, 5, 403–414.

Adams, C.C. 1902. Southeastern United States as a center of geographical distribution of fauna and flora. *Biological Bulletin of the Marine Biology Laboratory, Woods Hole*, 3, 115–131.

Ahyong, S.T., and M.C. Ebach. 1999. First occurrence of a subfossil stomatopod crustacean from Australia. *Alcheringa*, 25, 56–59.

Andersen, N.M. 1991. Cladistic biogeography of marine waterstriders (Insecta, Hemiptera) in the Indo-Pacific. *Australian Systematic Botany*, 4, 151–163.

Andersen, N.M. 1998. Marine water striders (Heteroptera, Gerromorpha) of the Indo-Pacific: Cladistic biogeography and Cenozoic palaeogeography. In R. Hall and J.D. Holloway (eds.), *Biogeography and geological evolution of SE Asia* (pp. 341–354). Backhuys Publishers, Leiden.

Anderson, D.L. 1984. The Earth as a planet: Paradigm and paradoxes. *Science*, 223, 347–355.

Anderson, S. 1994. Area and endemism. *Quarterly Review of Biology*, 69, 451–471.

Artedi, P. 1738. *Ichthyologia, sive Opera omnia piscibus scilicet: Bibliotheca ichthyologica. Philosophia ichthyologica. Genera piscium. Synonymia specierum. Descriptiones specierum. Omnia in hoc genere perfectiora, quam antea ulla. Posthuma vindicavit, recognovit, coaptavit et edidit Carolus Linnaeus, Med. Doct. Et Ac. Imper.* N. C. Wishoff, Leiden.

Avise, J.C. 2000. *Phylogeography: The history and formation of species.* Harvard University Press, Cambridge, Massachusetts.

Avise, J.C., J. Arnold, R.M. Ball, Jr, E. Bermingham, T. Lamb, J.E. Neigel, C.A. Reeb, and N.C. Saunders. 1987. Intraspecific phylogeography: the mitochondrial DNA bridge between population genetics and systematics. *Annual Review of Ecology and Systematics,*18, 489–522.

Axelius, B. 1991. Areas of distribution and areas of endemism. *Cladistics,* 7, 197–199.

Axelrod, D.I. 1972. Ocean-floor spreading in relation to ecosystematic problems. In R.T. Allen and F.C. James (eds.), *A symposium on ecosystematics,* conducted at the University of Arkansas, Fayetteville, Arkansas, April 21–22, 1971 (pp. 15–76). University of Arkansas Museum, Fayetteville.

Bernardin de Saint-Pierre, J.-H. 1804. *Etudes de la nature, volume* 3. Deterville, Paris.

Berra, T. 2001. *Freshwater fish distribution.* Academic Press, New York.

Bremer, K. 1992. Ancestral areas: A cladistic reinterpretation of the center of origin concept. *Systematic Biology,* 41, 436–445.

Bremer, K. 1995. Ancestral areas: Optimisation and probability. *Systematic Biology,* 44, 255–259.

Briggs, J.C. 1974. *Marine zoogeography.* McGraw-Hill, New York.

Briggs, J.C. 1995. *Global biogeography. Developments in palaeontology and stratigraphy 14.* Elsevier Science, Amsterdam.

Briggs, J.C. 1999. Coincident biogeographic patterns: Indo-West Pacific Ocean. *Evolution,* 53, 326–335.

Brooks, D.R. 1981. Hennig's parasitological method: A proposed solution. *Systematic Zoology,* 30, 229–249.

Brooks, D.R. 1985. Historical ecology: A new approach to studying the evolution of ecological associations. *Annals of the Missouri Botanical Garden,* 72, 660–680.

Brooks, D.R. 2005. Historical biogeography in the age of complexity: Expansion and integration. *Revista Mexicana de Biodiversidad,* 76, 79–94.

Brooks, D.R., A.P.G. Dowling, M.G.P. van Veller, and E.P. Hoberg. 2004. Ending a decade of deception: A valiant failure, a not-so-valiant failure, and a success story. *Cladistics,* 20, 32–46.

Brooks, D.R., and D.A. McLennan. 1991. *Phylogeny, ecology, and behavior.* University of Chicago Press, Chicago.

Brooks, D.R., and D.A. McLennan. 2002. *The nature of diversity: An evolutionary voyage of discovery.* University of Chicago Press, Chicago.

Brooks, D.R., M.G.P. van Veller, and D.A. McLennan. 2001. How to do BPA, really. *Journal of Biogeography,* 28, 343–358.

Browne, J. 1983. *The secular ark: Studies in the history of biogeography.* Yale University Press, New Haven, Connecticut.

Brundin, L. 1966. Transantarctic relationships and their significance as evidenced by chironomid midges. *Kungliga Svenska Vetenskapsakademiens Handlinger,* 11, 1–472.

Brundin, L. 1972. Phylogenetics and biogeography. *Systematic Zoology,* 21, 69–79.

Bueno-Hernández, A.A., and J.E. Llorente-Bousquets. 2006. The other face of Lyell: Historical biogeography in his *Principles of Geology. Journal of Biogeography,* 33, 549–559.

Buffon, G.L.L.C.D. 1766. *Histoire naturelle générale et particulière.* L'Imprimerie Royale, Paris.

Burbidge, N. 1960. The phytogeography of the Australian region. *Australian Journal of Botany,* 8, 75–211.

Camerini, J.R. 1993. Evolution, biogeography, and maps: An early history of Wallace's Line. *Isis,* 84, 700–727.

Campbell, S.W. 1999. Chemical weathering associated with tafoni at Papago Park, Central Arizona. *Earth Surface Processes and Landforms,* 24, 271–278.

Candolle, A.P.D. 1820. Essai élélmentare de géographie botanique. In *Dictionnaire des sciences naturelles.* F. Levrault, Strasbourg and Paris.

Candolle, A.L.P.P.D. 1855. *Géographie botanique raisonnée.* Masson, Paris.

Cárdenas, L., C.E. Hernández, E. Poulin, A. Magoulas, I. Kornfield, and F.P. Ojeda. 2005. Origin, diversification, and historical biogeography of the genus *Trachurus* (Perciformes, Carangidae). *Molecular Phylogenetics and Evolution,* 35, 496–507.

Carey, S.W. 1976. *The expanding Earth.* Elsevier, Amsterdam.

Carey, S.W. 1988. *Theories of the Earth and universe: A history of dogma in Earth sciences.* Stanford University Press, Palo Alto, California.

Carlquist, S. 1965. *Island life.* The Natural History Press, New York.

Carpenter, K.E., and V.G. Springer. 2005. The center of the center of marine shore fish biodiversity: The Philippine Islands. *Environmental Biology of Fishes,* 72, 467–480.

Cecca, F. 2002. *Palaeobiogeography of marine fossil invertebrates: Concepts and methods.* Taylor and Francis, New York.

Chakrabarty, P. 2004. Cichlid biogeography: Comment and review. *Fish and Fisheries,* 5, 97–119.

Christopherson, R.W. 2008. *Geosystems: An introduction to physical geography, seventh edition.* Prentice Hall, New Jersey.

Clements, F.E. 1905. *Research methods in ecology.* University Publishing Company, Lincoln, Nebraska.

Conaghan, P.J., I.W. Mountjoy, E.R. Edgecombe, J.R. Talent, and S.E. Owen. 1976. Nubrigyn algal reefs (Devonian), eastern Australia: Allochthonous blocks and megabreccias. *Geological Society of America Bulletin,* 87, 515–530.

Corner, E.J.H. 1963. *Ficus* in the Pacific region. In J.L. Gressitt (ed.), *Pacific Basin biogeography, a symposium* (pp. 233–245). Bishop Museum Press, Honolulu, Hawaii.

Cox, C.B. 2001. The biogeographic regions reconsidered. *Journal of Biogeography,* 28, 511–523.

Coyle, F.A. 1971. Systematics and natural history of the mygalomorph spider genus *Antrodiaetus* and related genera (Araneae: Antrodiaetidae). *Bulletin of the Museum of Comparative Zoology, Harvard University,* 141, 269–402.

Cracraft, J. 1979. Phylogenetic analysis, evolutionary models and paleontology. In J. Cracraft and N. Eldredge (eds.), *Phylogenetic analysis and paleontology* (pp. 7–39). Columbia University Press, New York.

Cracraft, J. 1985. Historical biogeography and patterns of differentiation within the South American avifauna: Areas of endemism. *Ornithological Monographs,* 36, 49–84.

Cracraft, J. 1991. Patterns of diversification within continental biotas: Hierarchical congruence among areas of endemism of Australian vertebrates. *Australian Systematic Botany,* 4, 211–227.

Craw, R.C. 1984. Biogeography and biogeographical principles. *New Zealand Entomologist,* 8, 49–52.

Craw, R.C. 1985. Classic problems of southern hemisphere biogeography re-examined: Panbiogeographic analysis of the New Zealand frog *Leiopelma,* the ratite birds and *Nothofagus. Zeitschrift zür Zoologische Systematik und Evolutionsforschung,* 23, 1–10.

Craw, R.C. 1988. Continuing the synthesis between panbiogeography, phylogenetic systematics and geology as illustrated by empirical studies on the biogeography of New Zealand and the Chatham Islands. *Systematic Zoology,* 37, 291–310.

Craw, R.C. 1989. New Zealand biogeography: A panbiogeographic approach. *New Zealand Journal of Zoology,* 16, 527–547.

Craw, R.C., J.R. Grehan, and M.J. Heads. 1999. *Panbiogeography: Tracking the history of life.* Oxford University Press, New York.

Crisci, J.V. 2001. The voice of historical biogeography. *Journal of Biogeography,* 28, 157–168.

Crisci, J.V. 2006. One-dimensional systematist: Perils in a time of steady progress. *Systematic Botany,* 31, 217–221.

Crisci, J.V., L. Katinas, and P. Posadas. 2003. *Historical biogeography: An introduction.* Harvard University Press, Cambridge, Massachusetts.

Crisci, J.V., O.E. Sala, L. Katinas, and P. Posadas. 2006. L.A.S. Johnson Review No. 4. Bridging historical and ecological approaches in biogeography. *Australian Systematic Botany,* 19, 1–10.

Crisp, M.D., S. Laffan, H.P. Linder, and A. Monro. 2001. Endemism in the Australian flora. *Journal of Biogeography,* 28, 183–198.

Croizat, L. 1952. *Manual of phytogeography or an account of plant dispersal throughout the world.* W. Junk, The Hague.

Croizat, L. 1958. *Panbiogeography or an introductory synthesis of zoogeography, phytogeography, and geology; with notes on evolution, systematics, ecology, anthropology, etc.* Published by the author, Caracas.

Croizat, L. 1961. *Principia botanica or beginnings of botany.* Published by the author, Caracas.

Croizat, L. 1964. *Space, time, form: The biological synthesis.* Published by the author, Caracas.

Croizat, L., G. Nelson, and D. E. Rosen. 1974. Centers of origin and related concepts. *Systematic Zoology,* 23, 265–287.

Cronk, Q. C. B. 1992. Relict floras of Atlantic islands: Patterns assessed. *Biological Journal of the Linnean Society,* 46, 91–103.

Cronk, Q. C. B. 1997. Islands: Stability, diversity, conservation. *Biodiversity and Conservation,* 6, 477–493.

Cronk, Q. C. B., M. Kiehn, W. L. Wagner, and J. F. Smith. 2005. Evolution of *Cyrtandra* (Gesneriaceae) in the Pacific Ocean: The origin of a supertramp clade. *American Journal of Botany,* 92, 1017–1024.

Crovello, T. J. 1981. Quantitative biogeography: An overview. *Taxon,* 30, 563–575.

Cunningham, C. W., and T. M. Collins. 1994. Developing model systems for molecular biogeography: Vicariance and interchange in marine invertebrates. In B. Schierwater, G. Streit, P. Wagner, and R. DeSalle (eds.), *Molecular ecology and evolution: Approaches and applications* (pp. 405–433). Birkhäuser Verlag, Basel.

Cutler, A. 2003. *The seashell on the mountaintop: How Nicholas Steno solved an ancient mystery and created a science of the Earth.* Dutton, New York.

Daeschler, E. B., N. H. Shubin, and F. A. Jenkins. 2006. A Devonian tetrapod-like fish and the evolution of the tetrapod body plan. *Nature,* 440, 757–763.

Dana, J. D. 1854. *A system of mineralogy, fourth edition.* Durrie and Peck, New Haven, Connecticut.

Dansereau, P. 1957. *Biogeography: An ecological perspective.* Ronald Press, New York.

Darlington, P. J., Jr. 1957. *Zoogeography: The geographical distribution of animals.* John Wiley and Sons, New York.

Darwin, C. 1859. *On the origin of species by means of natural selection, or the preservation of favoured races in the struggle for life.* John Murray, London.

Deo, A. J., and R. DeSalle. 2006. Nested areas of endemism analysis. *Journal of Biogeography,* 33, 1511–1526.

De Queiroz, A. 2005. The resurrection of oceanic dispersal in historical biogeography. *Trends in Ecology and Evolution,* 20, 68–73.

De Queiroz, K. 2000. The definitions of taxon names: A reply to Stuessy. *Taxon,* 49, 533–536.

De Queiroz, K., and P. D. Cantino. 2001. Phylogenetic nomenclature and the PhyloCode. *Bulletin of Zoological Nomenclature,* 58, 254–271.

De Queiroz, K., and J. Gauthier. 1990. Phylogeny as a central principle in taxonomy: Phylogenetic definitions of some taxon names. *Systematic Zoology,* 39, 307–322.

De Queiroz, K., and J. Gauthier. 1992. Phylogenetic taxonomy. *Annual Review of Ecology and Systematics,* 23, 449–480.

Dickerson, R. E., in collaboration with E. D. Merrill, R. C. McGregor, W. Schultze, E. H. Taylor, and A. W. C. T. Herre. 1928. Distribution of life in the Philippines. Monographs of the Bureau of Science, Manila, Philippine Islands. Monograph 21, 1–322.

Dietz, R.S., and J.C. Holden. 1970. Reconstruction of Pangaea: Breakup and dispersion of continents, Permian to present. *Journal of Geophysical Research*, 75, 4939–4956.

Dobson, J.E. 1992. Spatial logic in paleogeography and the explanation of continental drift. *Annals of the Association of American Geographers*, 82, 187–206.

Domínguez, M.C., S. Roig-Juñent, J.J. Tassin, F.C. Ocampo, and G.E. Flores. 2006. Areas of endemism of the Patagonian steppe: An approach based on insect distributional patterns using endemicity analysis. *Journal of Biogeography*, 33, 1527–1537.

Donoghue, M.J., and B.R. Moore. 2003. Toward an integrative historical biogeography. *Integrative and Comparative Biology*, 43, 261–270.

Ducasse, J., N. Cao, and R. Zaragüeta-Bagils. 2007. *Nelson05: Three-item analysis software*. Laboratoire Informatique et Systématique, UPMC Univ Paris 06, et Muséum national d'Histoire naturelle, Paris.

Du Toit, A.L. 1937. *Our wandering continents*. Oliver and Boyd, Edinburgh.

Ebach, M.C. 1999. Paralogy and the centre of origin concept. *Cladistics*, 15, 387–391.

Ebach, M.C. 2003. Area cladistics. *Biologist*, 50, 169–172.

Ebach, M.C., and G.D. Edgecombe. 1999. The Devonian trilobite *Cordania* from Australia. *Journal of Palaeontology*, 73, 431–436.

Ebach, M.C., and G.D. Edgecombe. 2001. Cladistic biogeography: Component-based methods and paleontological application. In J.M. Adrain, G.D. Edgecombe, and B.S. Lieberman (eds.), *Fossils, phylogeny and form: An analytical approach* (pp. 235–289) Plenum, New York.

Ebach, M.C., and D.F. Goujet. 2006. The first biogeographical map. *Journal of Biogeography*, 33, 1–9.

Ebach, M.C., and C.J. Humphries. 2002. Cladistic biogeography and the art of discovery. *Journal of Biogeography*, 20, 427–444.

Ebach, M.C., C.J. Humphries, R.A. Newman, D.M. Williams, and S.A. Walsh. 2005. Assumption 2: Opaque to intuition? *Journal of Biogeography*, 32, 781–787.

Ebach, M.C., C.J. Humphries, and D.M. Williams. 2003. Phylogenetic biogeography deconstructed. *Journal of Biogeography*, 30, 1285–1296.

Ebach, M.C., and J.J. Morrone. 2005. Forum on historical biogeography: What is cladistic biogeography? *Journal of Biogeography*, 32, 2179–2187.

Ebach, M.C., J.J. Morrone, L.R. Parenti, and Á.L. Viloria. 2008. International Code of Area Nomenclature. *Journal of Biogeography*, 35, 1153–1157.

Egerton, F.N. 1984. [Review of] Browne, J. The secular ark: Studies in the history of biogeography. Yale University Press, New Haven. *Isis*, 75, 405–406.

Ekman, S. 1935. *Tiergeographie des meeres*. Akademische Verlagsgesellschaft, Leipzig.

Ekman, S. 1953. *Zoogeography of the sea*. Sidgwick and Jackson, London.

Elith, J., C.H. Graham, R.P. Anderson, M. Dudik, S. Ferrier, A. Guisan, R.J. Hijmans, F. Huettmann, J.R. Leathwick, A. Lehmann, J. Li, L.G. Lohmann, B.A. Loiselle, G. Manion, C. Moritz, M. Nakamura, Y. Nakazawa, J.M. Overton, A.T. Peterson, S.J. Phillips, K.S. Richardson, R. Scachetti-Pereira,

R. E. Schapire, J. Soberón, S. Williams, M. S. Wisz, and N. E. Zimmermann. 2006. Novel methods improve prediction of species' distributions from occurrence data. *Ecography,* 29, 129–151.

Enghoff, H. 1996. Historical biogeography of the holarctic: Area relationships, and dispersal of marine mammals. *Cladistics,* 11, 223–263.

Engler, A. 1879. *Versuch einer entwicklungsgeschichte der pflanzewelt insbesondere der florengebiete siet der tertiärperiode.* Engelmann, Leipzig.

Engler, A. 1882. *Versuch einer entwicklungsgeschichte der pflanzewelt.* Engelmann, Leipzig.

Escalante, T., G. Rodríguez, N. Cao, M. C. Ebach, and J. J. Morrone. 2007. Cladistic biogeographic analysis suggests a Caribbean diversification prior to the Great American Biotic Interchange and the Mexican Transition Zone. *Naturwissenschaften,* 94, 561–565.

Everhart, M. J. 2005. *Oceans of Kansas: A natural history of the western interior sea.* Indiana University Press, Bloomington.

Farias, I. P., G. Orti, and A. Meyer. 2000. Total evidence: Molecules, morphology, and the phylogenetics of cichlid fishes. *Journal of Experimental Zoology,* 288, 76–92.

Fattorini, S. 2007. Levels of endemism are not necessarily biased by the co-presence of species with different size ranges: A case study of Vilenkin and Chikatunov's models. *Journal of Biogeography,* 34, 994–1007.

Fattorini, S. 2008. Hovenkamp's ostracized vicariance analysis: Testing new methods of historical biogeography. *Cladistics,* 24, 611–622.

Fauchald, K. 1984. Polychaete distribution patterns, or: Can animals with Palaeozoic cousins show large-scale geographical patterns. In P. A. Hutchings (ed.), *Proceedings of the First International Polychaete Conference,* Sydney, Australia, July, 1983 (pp. 1–6). The Linnean Society of New South Wales, Sydney.

Felsenstein, J. 2003. *Inferring phylogenies.* Sinauer Associates Inc., Sunderland, Massachusetts.

Ferraiolo, J. A. 1982. A systematic classification of nonsilicate minerals. *Bulletin of the American Museum of Natural History,* 172, 1–237.

Ferris, V. R. 1980. A science in search of a paradigm? Review of the symposium "Vicariance Biogeography: A Critique." *Systematic Zoology,* 29, 67–76.

Fisher, D. C. 1994. Stratocladistics: Morphological and temporal patterns and their relation to phylogenetic process. In G. Grande and O. Rieppel (eds.), *Interpreting the hierarchy of nature: From systematic patterns to evolutionary process theories* (pp. 133–171). Academic Press, NewYork.

Fisher, D. C. 2008. Stratocladistics: Integrating temporal data and character data in phylogenetic inference. *Annual Review of Ecology, Evolution, and Systematics,* 39, 365–385.

Fitch, W. M. 1970. Distinguishing homologous from analogous proteins. *Systematic Zoology,* 19, 99–113.

Foin, T. C. 1976. Plate tectonics and the biogeography of the Cypraeidae (Mollusca: Gastropoda). *Journal of Biogeography,* 3, 19–34.

Forbes, E. 1846. On the connexion between the distribution of the existing fauna and flora of the British Isles and the geological changes which have affected their area, especially during the epoch of the Northern Drift. *Memoirs of the Geological Survey of Great Britain,* 1, 336–432.

Forbes, E. 1854 (1856). Map of the distribution of marine life, illustrated chiefly by fishes, molluscs and radiata; showing also the extent and limits of the homoizoic belts. In A. K. Johnston (ed.), *The physical atlas of natural phenomena* (Plate 31). William Blackwood and Sons, Edinburgh.

Forey, P.L. 1982. Neontological analysis versus palaeontological stories. In K.A. Joysey and A.E. Friday (eds.), *Problems of phylogenetic reconstruction* (pp. 197–234). Systematics Association Special Volume 21, Academic Press, London.

Forey, P.L. 1998. A home from home for coelacanths. *Nature,* 395, 319–320.

Forey, P.L. 2001. The PhyloCode: Description and commentary. *Bulletin of Zoological Nomenclature,* 58, 81–96.

Fortey, R. 1996. *Life: A natural history of the first four billion years of life on Earth.* Alfred Knopf, New York.

Fox, D.L., D.C. Fisher, and L.R. Leighton. 1999. Reconstructing phylogeny with and without temporal data. *Science,* 284, 1816–1819.

Gee, H. 2000. *Deep time: Cladistics, the revolution in evolution.* Fourth Estate, London.

Gemmill, C.E.C., G.J. Allan, W.L. Wagner, and E.A. Zimmer. 2002. Evolution of insular Pacific *Pittosporum* (Pittosporaceae): Origin of the Hawaiian radiation. *Molecular Phylogenetics and Evolution,* 22, 31–42.

Ghiselin, M.T. 2006. The failure of morphology to contribute to the modern synthesis. *Theory in Biosciences,* 124, 309–316.

Ghosh, S. C. 2002. The Raniganj Coal Basin: An example of an Indian Gondwana Rift. *Sedimentary Geology,* 147, 155–176.

Gingerich, P.D. 1979. The stratophenetic approach to phylogeny resonstruction in vertebrate paleontology. In J. Cracraft and N. Eldredge (eds.), *Phylogenetic analysis and paleontology* (pp. 41–77). Columbia University Press, New York.

Giokas, S., and S. Sfenthourakis. 2008. An improved method for the identification of areas of endemism using species co-occurrences. *Journal of Biogeography,* 35, 893–902.

Giraud-Soulavie, J.-L. 1770–1784. *Histoire naturelle de la France méridionale.* Nismes, Paris.

Glasby, C.J. 2005. Polychaete distribution patterns revisited: An historical explanation. *Marine Ecology,* 26, 235–245.

Goethe, J.W. von. 1995. *Scientific studies.* Princeton University Press, New Jersey.

Goethe, J.W. von. 2006. *Theory of colours.* Dover Publications, Inc., Mineola, New York.

Goldenfeld, N., and C. Woese. 2007. Biology's next revolution. *Nature,* 445, 369.

Golikov, A.N., M.A. Dolgolenko, N.V. Maximovich, and O.A. Scarlato. 1990. Theoretical approaches to marine biogeography. *Marine Ecology Progress Series,* 63, 289–301.

Good, R. 1964. *The Geography of flowering plants.* Longman, London.

Grande, L. 1985. The use of paleontology in systematics and biogeography, and a time control refinement for historical biogeography. *Paleobiology,* 11, 234–243.

Grande, L., and W.E. Bemis. 1991. Osteology and phylogenetic relationships of fossil and recent paddlefishes (Polyodontidae) with comments on the interrelationships of Acipenseriformes. *Journal of Vertebrate Paleontolology* 11 (Memoir 1), 1–121.

Grande, L., J. Fan, Y. Yabumoto, and W.E. Bemis. 2002. †*Protosephurus liui*, a well-preserved primitive paddlefish (Acipenseriformes: Polyodontidae) from the early Cretaceous of China. *Journal of Vertebrate Paleontolology*, 22, 209–237.

Grehan, J.R. 2006. A brief look at Pacific biogeography: The trans-oceanic travels of *Microseris* (Angiosperms:Asteraceae). In M.C. Ebach and R.S. Tangney (eds.), *Biogeography in a changing world* (pp. 83–84). Fifth Biennial Conference of the Systematics Association, UK, CRC Press, Boca Raton, Florida.

Gressitt, J.L. 1961. Problems in the zoogeography of Pacific and Antarctic insects. *Pacific Insects Monograph*, 2, 1–94.

Gressitt, J.L. (ed.). 1963. *Pacific basin biogeography: A symposium*. Bishop Museum Press, Honolulu, Hawaii.

Griswold, C.E. 1991. Cladistic biogeography of Afromontane spiders. *Australian Systematic Botany*, 4, 73–89.

Grotzinger, J., T.H. Jordan, F. Press, and R. Siever. 2006. *Understanding Earth, fifth edition*. W.H. Freeman, New York.

Haeckel, E. 1866. *Generelle morphologie der organismen*. G. Reimer, Berlin.

Haffer, J. 1969. Speciation in Amazonian forest birds. *Science*, 165, 131–137.

Hall, R. 1996. Reconstructing Cenozoic Southeast Asia. In R. Hall and D. Blundell (eds.), *Tectonic evolution of Southeast Asia* (pp. 153–184). Geological Society Special Publication no. 106. The Geological Society, London.

Hall, R. 2002. Cenozoic geological and plate tectonic evolution of SE Asia and the SW Pacific: Computer-based reconstructions and animations. *Journal of Asian Earth Sciences*, 20, 353–434.

Hallam, A. 1973. *A revolution in the earth sciences: From continental drift to plate tectonics*. Clarendon Press, Oxford.

Harold, A.S., and R.D. Mooi. 1994. Areas of endemism: Definition and recognition criteria. *Systematic Biology*, 43, 261–266.

Hausdorf, B. 1998. Weighted ancestral area analysis and a solution of the redundant distribution problem. *Systematic Biology*, 47, 445–456.

Hausdorf, B. 2002. Units in biogeography. *Systematic Biology*, 51, 648–652.

Hawthorne, F.C. 1985. Towards a structural classification of minerals: The V1MIVT2φn minerals. *American Mineralogist*, 70, 455–473.

Heads, M.J. 2003. Ericaceae in Malesia: Vicariance biogeography, terrane tectonics and ecology. *Telopea*, 10, 311–449.

Heads, M.J. 2005a. Dating nodes on molecular phylogenies: A critique of molecular biogeography. *Cladistics*, 21, 62–78.

Heads, M.J. 2005b. Towards a panbiogeography of the seas. *Biological Journal of the Linnean Society*, 84, 675–723.

Hendrickson, D. 1986. Congruence of bolitoglossine biogeography and phylogeny with geological history: Paleotransport on displaced suspect terranes. *Cladistics*, 2, 113–129.

Hennig, W. 1950. *Grundzüge einer theorie der phylogenetischen systematik.* Deutscher Zentralverlag, Berlin.

Hennig, W. 1965. Phylogenetic systematics. *Annual Review of Entomology,* 10, 97–116.

Hennig, W. 1966. *Phylogenetic systematics.* University of Illinois Press, Urbana.

Hess, H.H. 1962. History of ocean basins. In A.E. Engel, H.L. James and B.F Cohen (eds.), *Petrologic studies: A volume in honour of A.F. Buddington* (pp. 599–620). Geological Society of America, Colorado.

Hinz, P.-A. 1989. L'endémisme: 1. Concepts généraux. *Saussurea,* 20, 145–168.

Hooker, J.D. 1853. *The botany of the Antarctic voyage of H.M. discovery ships* Erebus *and* Terror *in the years 1839–1843, volume II, Flora Novae Zelandiae, part 1, Flowering plants.* Lovell Reeve, London.

Hooker, J.D. 1860. *Botany of the Antarctic voyage of H.M. discovery ships* Erebus *and* Terror *in the years 1839–1843, volume III, Flora Tasmaniae.* Lovell Reeve, London.

Hooper, M., and C. Cody. 1996. *The pebble in my pocket: A history of our Earth.* Viking Press, New York.

Hovenkamp, P. 1997. Vicariance events, not areas, should be used in biogeographical analysis. *Cladistics,* 13, 67–79.

Hovenkamp, P. 2001. A direct method for the analysis of vicariance patterns. *Cladistics,* 17, 260–265.

Hull, D.L. 1988. *Science as process: An evolutionary account of the social and conceptual development of science.* University of Chicago Press, Chicago.

Humboldt, A.V., and A.J.A. Bonpland. 1805. *Essai sur la géographie des plantes: Accompagné d'un tableau physique des régions équinoxiales.* Levrault, Schoell et Compagnie, Paris.

Humphries, C.J. 2004. From dispersal to geographic congruence: Comments on cladistic biogeography in the twentieth century. In D.M. Williams and P.L. Forey (eds.), *Milestones in systematics* (225–260). Systematics Association special volume 67. CRC Press, Boca Raton, Florida.

Humphries, C.J., and L.R. Parenti. 1986. *Cladistic biogeography.* Clarendon Press, Oxford.

Humphries, C.J., and L.R. Parenti. 1999. *Cladistic biogeography, second edition: Interpreting patterns of plant and animal distributions.* Oxford University Press, Oxford.

Hunn, C.A., and P. Upchurch. 2001. The importance of time/space in diagnosing the causality of phylogenetic events: Towards a "chronobiogeographical" paradigm? *Systematic Biology,* 50, 391–407.

Hüpsch, J.W.K.A.von. 1764. *Physikalische abhandlung von der vormaligen verknüpfung und absonderung der alten und neuen welt, und der Bevölkerung Westindies: Nebst einer physikalischen untersuchung von dem ursprung der seen.* Johan Heintrich Hartz, Cöln am Rheine.

Huxley, T.H. 1894. *Darwiniana: Essays by Thomas H. Huxley.* D. Apptleton and Company, New York.

Jablonski, D., K. Flessa, and J.W. Valentine. 1985. Biogeography and paleobiology. *Paleobiology,* 11, 75–90.

Janvier, P. 2007. Living primitive fishes and fishes from deep time. In D.J. McKenzie, A.P. Farrell, and C.J. Brauner (eds.), *Primitive fishes. Fish physiology, volume* 26 (pp. 1–51). Academic Press, New York.

Kay, E.A. 1980. *Little worlds of the Pacific: An essay on Pacific Basin biogeography.* University of Hawaii, Honolulu.

Kearey, P., and F. Vine. 1996. *Global tectonics, second edition.* Blackwell Science, London.

Keast, A., and S.E. Miller (eds.). 1996. *The origin and evolution of Pacific Island biotas, New Guinea to Eastern Polynesia: Patterns and processes.* SPB Academic Publishing, Amsterdam.

Keppel, G., P.D. Hodgskiss, and G.M. Plunkett. 2008. Cycads in the insular South-west Pacific: Dispersal or vicariance? *Journal of Biogeography,* 35, 1004–1015.

Kinch, M.P. 1980. Geographical distribution and the origin of life: The development of early nineteenth century British explanations. *Journal of the History of Biology,* 13, 91–119.

Kitching, I.J., P. L. Forey, C.J. Humphries, and D.M. Williams.1998. *Cladistics, second edition: The theory and practice of parsimony analysis.* Oxford University Press, Oxford.

Knoll, A.H. 2003. *Life on a young planet: The first three billion years of evolution on Earth.* Princeton University Press, Princeton, New Jersey.

Krell, F.-T., and P.S. Cranston. 2004. Which side of the tree is more basal? *Systematic Entomology,* 29, 279–281.

Ladiges, P.Y. 1998. Biogeography after Burbidge. *Australian Systematic Botany,* 11, 231–242.

Ladiges, P.Y., J. Kellermann, G. Nelson, C.J. Humphries, and F. Udovicic. 2005. Historical biogeography of Australian Rhamnaceae, tribe Pomaderreae. *Journal of Biogeography,* 32, 1909–1919.

Ladiges, P.Y., S.M. Prober, and G. Nelson. 1992. Cladistic and biogeographic analysis of the 'blue-ash' eucalypts. *Cladistics,* 8, 103–124.

Laffan, S.W., and M.D. Crisp. 2003. Assessing endemism at multiple spatial scales, with an example from the Australian vascular flora. *Journal of Biogeography,* 30, 511–520.

Lamarck, J.B.P.A.D.M.D. 1778. *Flore françoise, ou, description succinte de toutes les plantes qui croissent naturellement en France: Disposée selon une nouvelle méthode d'analyse, et à laquelle on a joint la citation de leurs vertus les moins équivoques en médicine, et de leur utilité dans les arts.* L'Imprimerie Royale, Paris.

Lamarck, J.B.P.A.D.M.D, and A.P.D. Candolle. 1805. *Flore française, ou descriptions succinctes de toutes les plantes qui croissent naturellement en France: Disposées selon une nouvelle méthode d'analyse, et précédées par un exposé des principes élémentaires de la botanique.* Desray, Paris.

Latiolais, J.M., M.S. Taylor, K. Roy, and M.E. Hellberg. 2006. A molecular phylogenetic analysis of strombid gastropod morphological diversity. *Molecular Phylogenetics and Evolution,* 41, 436–444.

Lees, D.C., R.A. Fortey, and R. M. Cocks. 2002. Quantifying paleogeography using biogeography: A test case for the Ordovician and Silurian of Avalonia based on brachiopods and trilobites. *Paleobiology,* 28, 343–363.

Leviton, A.E., and M.L. Aldrich (eds.). 1986. *Plate Tectonics and Biogeography. Earth Sciences History, Journal of the History of the Earth Sciences Society,* 4(2), 91–196.

Lieberman, B.S. 2000. *Paleobiogeography: Using fossils to study global change, plate tectonics, and evolution.* Kluwer Academic/Plenum Publishers, New York.

Lieberman, B.S., and N. Eldredge. 1996. Trilobite biogeography in the Middle Devonian: Geological processes and analytical methods. *Paleobiology,* 22, 66–79.

Linnaeus, C. 1735. *Systema naturae, sive regna tria naturae sytematice proposita per classes, ordines, generes et species.* deGroot, Leiden.

Linnaeus, C. 1758. *Systema naturae per regna tria naturae, secundum classes, ordines, genera, species, cumcharacteribus, differentiis, synonymis, locis.* Laurentii Salvii, Stockholm.

Lomolino, M.V., and L.R. Heaney (eds.). 2004. *Frontiers of biogeography: New directions in the geography of nature.* Sinauer Associates Inc., Sunderland, Massachusetts.

Lomolino, M.V., B.R. Riddle, and J.H. Brown. 2005. *Biogeography.* Sinauer Associates Inc., Sunderland, Massachusetts.

Lomolino, M.V., D.F. Sax, and J.H. Brown (eds.). 2004. *Foundations of biogeography: Classic papers with commentaries.* University of Chicago Press, Chicago.

Longhurst, A.R. 1998. *Ecological geography of the sea.* Academic Press, New York.

Lovelock, J. 1979. *Gaia: A new look at life on Earth.* Oxford University Press, Oxford.

Lyell, C. 1830–1833. *The principles of geology: Being an attempt to explain the former changes of the Earth's surface, by reference to causes now in operation* John Murray, London.

MacArthur, R.H., and E.O. Wilson. 1963. An equilibrium theory of insular biogeography. *Evolution,* 17, 373–387.

MacArthur, R.H., and E.O. Wilson. 1967. *The theory of island biogeography.* Princeton University Press, Princeton, New Jersey.

Marshall, C.J., and J.K. Liebherr. 2000. Cladistic biogeography of the Mexican Transition Zone. *Journal of Biogeography,* 27, 203–216.

Mast, A.R., and R. Nyffeler. 2003. Using a null model to recognize significant co-occurrence prior to identifying candidate areas of endemism. *Systematic Biology,* 52, 271–280.

Mayden, R.L. 1989. Phylogenetic studies of North American minnows with emphasis on the genus *Cyprinella*, new genus, new status (Teleostei: Cypriniformes). *University of Kansas Museum of Natural History Miscellaneous Publications,* 80, 1–189.

Mayr, E. 1942. *Systematics and the origin of species.* Dover Press, New York.

Mayr, E. 1946. History of the North American bird fauna. *The Wilson Bulletin,* 58, 3–41.

Mayr, E. 1974. Cladistic analysis or cladistic classification? *Zeitschrift für Zoologische Systematik und Evolutionsforschung,* 12, 94–128.

Mayr, E. 1982. *The growth of biological thought: Diversity, evolution and inheritance.* Belknap Press, Harvard University Press, Cambridge, Massachusetts.

Mayr, E., and C.B. Rosen. 1956. Geographic variation and hybridization in populations of Bahama snails (*Cerion*). *American Museum Novitates,* 1806, 1–48.

McCarthy, D. 2003. The trans-Pacific zipper effect: Disjunct sister taxa and matching geological outlines that link the Pacific margins. *Journal of Biogeography,* 30, 1545–1561.

McCarthy, D. 2005. Biogeographical and geological evidence for a smaller, completely-enclosed Pacific Basin in the Late Cretaceous. *Journal of Biogeography,* 32, 2161–2177.

McCarthy, D., M.C. Ebach, J.J. Morrone, and L.R. Parenti. 2007. An alternative Gondwana: Biota links South America, New Zealand and Australia. *Biogeografía,* 2, 1–11.

McCoy, E.D., and K. Heck Jr. 1976. Biogeography of corals, seagrasses, and mangroves: An alternative to the center of origin concept. *Systematic Zoology,* 25, 201–210.

McDowall, R., and W. Fulton. 1996. Family Galaxiidae: Galaxiids. In R. McDowall (ed.), *Freshwater fishes of south-eastern Australia* (pp. 52–80). Reed Books, Chatswood.

Meacham, C.A. 1981. A probability measure for character compatibility. *Mathematical Biosciences,* 57, 1–18.

Meacham, C.A., and G.F. Estabrook. 1985. Compatibility methods in systematics. *Annual Review of Ecology and Systematics,* 16, 431–446.

Merriam, C.H. 1892. The geographical distribution of life in North America with special reference to the Mammalia. *Proceedings of the Biological Society of Washington,* 7, 1–64.

Merriam, C.H. 1899. *Results of a biological survey of Mount Shasta, California.* United States Department of Agriculture, Division of Biological Survey, North American Fauna, 16. Washington, D.C.

Metcalfe, I. 2001. Palaeozoic and Mesozoic tectonic evolution and biogeography of SE Asia-Australasia. In I. Metcalfe, J.M.B. Smith, M. Morwood, and I. Davidson (eds.), *Faunal and floral migrations and evolution in SE Asia–Australasia* (pp. 15–34). Balkema, Lisse.

Meyer, C.P., J.B. Geller, and G. Paulay. 2005. Fine scale endemism on coral reefs: Archipelagic differentiation in turbinid gastropods. *Evolution,* 59,113–125.

Michaux, B. 1996. The origin of southwest Sulawesi and other Indonesian terranes: a biological view. *Palaeogeography, Palaeoclimatology, Palaeoecology,* 122, 167–183.

Mickevich, M.F., and N.I. Platnick. 1989. On the information content of classification. *Cladistics,* 5, 33–47.

Miller, C.B. (ed.) 1974. *The biology of the oceanic Pacific. Proceedings of the thirty-third annual biology colloquium.* Oregon State University Press, Corvallis.

Miller, P.J. 1990. The endurance of endemism: The Mediterranean freshwater gobies and their prospects for survival. *Journal of Fish Biology,* 37, 145–156.

Montanucci, R.R. 1987. A phylogenetic study of the horned lizards, genus *Phrynosoma*, based on skeletal and external morphology. *Natural History Museum of Los Angeles County, Science Series,* 390, 1–36.

Mooi, R.D., and A.C. Gill. 2004. Notograptidae, sister to *Acanthoplesiops* Regan (Teleostei: Plesiopidae: Acanthoclininae), with comments on biogeography, diet and morphological convergence with Congrogadinae (Teleostei: Pseudochromidae). *Zoological Journal of the Linnean Society,* 141, 179–205.

Moreira-Muñoz, A. 2007. The Austral floristic realm revisited. *Journal of Biogeography,* 34, 1649–1660.

Morrone, J.J. 1994. On the identification of areas of endemism. *Systematic Biology,* 43, 438–441.

Morrone, J.J. 2001a. Homology, biogeography and areas of endemism. *Diversity and Distributions,* 7, 297–300.

Morrone, J.J. 2001b. *Sistemática, biogeografía, evolucíon: Los patrones de la biodiversidad en tiempo-espacio.* Las Prensas de Ciencias, Facultad de Ciencias, Universidad Nacional Autónoma de México, México, D.F.

Morrone, J.J. 2002. Biogeographical regions under track and cladistic scrutiny. *Journal of Biogeography,* 29, 149–152.

Morrone, J.J., and J.M. Carpenter. 1994. In search of a method for cladistic biogeography: An empirical comparison of component analysis, Brooks parsimony analysis, and three-area statements. *Cladistics,* 10, 99–153.

Morrone, J.J., and J.V. Crisci. 1995. Historical biogeography: Introduction to methods. *Annual Review of Ecology and Systematics,* 26, 373–401.

Morrone, J.J., and J.E. Llorente-Bousquets. 2003. *Una perspectiva latinoamericana de la biogeografía.* Las Prensas de Ciencias, Facultad de Ciencias, Universidad Nacional Autónoma de México, México, D.F.

Müller, G.H. 1995. *Friedrich Ratzel (1844–1904): Naturwissenschaftler, geograph, gelehrter; neue studien zu leben und werk und sein konzept der "allgemeinen biogeographie."* GNT-Verlag, Diepholz.

Munsell, A.H. 1905. A color notation. Ellis, Boston.

Murphy, W.J., and G.E. Collier. 1997. A molecular phylogeny for aplocheiloid fishes (Atherinomorpha, Cyprinodontiformes): The role of vicariance and the origins of annualism. *Molecular Biology and Evolution,* 14, 790–799.

Murray, A. 1866. *The geographical distribution of mammals.* Day and Son, London.

Myers, N., R.A. Mittermeier, C.G. Mittermeier, G.A.B. Da Fonseca, and J. Kent. 2000. Biodiversity hotspots for conservation priorities. *Nature,* 403, 853–858.

Naef, A. 1919. *Idealistische morphologie und phylogenetik (zur methodik der systematischen morphologie).* Gustav Fischer, Jena.

Nelson, G.J. 1978. From Candolle to Croizat: Comments on the history of biogeography. *Journal of the History of Biology,* 11, 296–305.

Nelson, G.J. 1979. Cladistic analysis and synthesis: Principles and definitions, with a historical note on Adanson's *Familles des Plantes. Systematic Zoology,* 28, 1–21.

Nelson, G. 1984. Cladistics and biogeography. In T. Duncan and T.F. Stuessy (eds.), *Cladistics: Perspectives on the reconstruction of evolutionary history* (pp. 273–293). Columbia University Press, New York.

Nelson, G. 1996. Nullius in verba. *Journal of Comparative Biology,* 1(3/4), 141–152.

Nelson, G., J.W. Atz, K.D. Kallman, and C.L. Smith. 1987. Donn Eric Rosen 1929–1986. *Copeia,* 1987, 541–547.

Nelson, G., and P.Y. Ladiges. 1991a. TAS: MSDOS program for cladistic biogeography. Published by the authors, New York and Melbourne.

Nelson, G., and P.Y. Ladiges. 1991b. Standard assumptions for biogeographic analysis. *Australian Systematic Botany,* 4, 41–58.

Nelson, G., and P.Y. Ladiges. 1991c. Three-area statements: Standard assumptions for biogeographic analysis. *Systematic Zoology,* 40, 470–485.

Nelson, G., and P.Y. Ladiges. 1994. TASS versions1.4, 1.5. Three area subtrees. Published by the authors, New York and Melbourne.

Nelson, G., and P.Y. Ladiges. 1995. TASS version 2.0. Three area subtrees. Published by the authors, New York and Melbourne.

Nelson, G., and P.Y. Ladiges. 1996. Paralogy in cladistic biogeography and analysis of paralogy-free subtrees. *American Museum Novitates,* 3167, 1–58.

Nelson, G., and N.I. Platnick. 1981. *Systematics and biogeography: Cladistics and vicariance.* Columbia University Press, New York.

Nelson, G., and N.I. Platnick. 1984. *Biogeography.* Carolina Biological Supply Company, Burlington, North Carolina.

Nelson, G., and N.I. Platnick. 1991. Three-taxon statements: A more precise use of parsimony? *Cladistics,* 7, 351–366.

Nelson, G., and D.E. Rosen. 1981. *Vicariance biogeography, a critique.* Columbia University Press, New York.

Newbigin, M.I. 1950. *Plant and animal geography, third edition.* E.P. Dutton, New York.

Nield, T. 2007. *Supercontinent: Ten billion years in the life of our planet.* Harvard University Press, Cambridge, Massachusetts.

Nihei, S. 2006. Misconceptions about parsimony analysis of endemicity. *Journal of Biogeography,* 33, 2099–2106.

Nudds, J.R., and P.A. Selden. 2008. *Fossil ecosystems of North America.* Manson, London.

Nur, A., and Z. Ben-Avraham. 1977. Lost Pacifica continent. *Nature,* 270, 41–43.

Olson, D.M., E. Dinerstein, E.D. Wikramanayake, N.D. Burgess, G.V. Powell, E.C. Underwood, J.A. D'amico, I. Itoua, H.E. Strand, J.C. Morrison, C.J. Loucks, T.F. Allnutt, T.H. Ricketts, Y. Kura, J.F. Lamoreux, W.W. Wettengel, P. Hedao, and K.R. Kassen. 2001. Terrestrial ecoregions of the world: A new map of life on Earth. *Bioscience,* 51, 933–938.

Owen, H.G. 1976. Continental displacement and expansion of the Earth during the Mesozoic and Cenozoic. *Philosophical Transactions of the Royal Society,* 281, 223–290.

Page, R.D.M. 1989. New Zealand and the new biogeography. *New Zealand Journal of Zoology,* 16, 471–483.

Page, R.D.M. 1990. Component analysis: A valiant failure? *Cladistics,* 6, 119–136.

Page, R.D.M. 1993. COMPONENT, version 2.0. The Natural History Museum, London.

Papavero, N., D. Martins Teixera, and J. Llorente-Bousquets. 1997. *Història da biogeografia no período prévolutivo.* Plêiade, FAPESP, São Paulo.

Papavero, N., M. Souto Couri, W.L. Overhal, and D.M. Teixeira. 2003. The 'physical dissertation on the former union and separation of the Old and New Worlds, and the peopling of the West Indies' (1764): The first proposal of a 'Pangaea supercontinent.' In J.J. Morrone and J. Llorente Bousquets (eds.), *Una perspectiva latinoamericana de la biogeografía* (pp. 9–18). Las Prensas de Ciencias, Facultad de Ciencias, Universidad Nacional Autónoma de México, México, D.F.

Parenti, L.R. 1981. A phylogenetic and biogeographic analysis of cyprinodontiform fishes. *Bulletin of the American Museum of Natural History,* 168, 335–557.

Parenti, L.R. 1989. A phylogenetic revision of the phallostethid fishes (Atherinomorpha, Phallostethidae). *Proceedings of the California Academy of Sciences,* 46, 243–277.

Parenti, L.R. 1991. Ocean basins and the biogeography of freshwater fishes. *Australian Systematic Botany,* 4, 137–149.

Parenti, L.R. 2005. The phylogeny of atherinomorphs: Evolution of a novel fish reproductive system. In M.-C. Uribe and H.J. Grier (eds.), *Viviparous fishes* (pp. 13–30). New Life Publications Inc., Homestead, Florida.

Parenti, L.R. 2006. Common cause and historical biogeography. In M.C. Ebach and R.S. Tangney (eds.), *Biogeography in a changing world* (pp. 61–82). Fifth Biennial Conference of the Systematics Association, UK. CRC Press, Boca Raton, Florida.

Parenti, L.R. 2008. Life history patterns and biogeography: an interpretation of diadromy in fishes. *Annals of the Missouri Botanical Garden,* 95, 232–247.

Patterson, C. 1982. Morphological characters and homology. In K.A. Joysey and A. Friday (eds.), *Problems of phylogenetic reconstruction* (pp. 21–74). Academic Press, London.

Patterson, C. 1983. Aims and methods in biogeography. In R.W. Sims, J.H. Price, and P.E.S. Whalley (eds.), *Evolution, time and space: The emergence of the biosphere* (pp. 1–28). Systematics Association Special volume 23, Academic Press, New York.

Pelletier, J.D. 1999. Species-area relation and self-similarity in a biogeographical model of speciation and extinction. *Physical Review Letters,* 82, 1983–1986.

Peterson, A.T., J. Soberón, and V. Sánchez-Cordero. 1999. Conservatism of ecological niches in evolutionary time. *Science,* 285, 1265–1267.

Pielou, E. 1979. *Biogeography.* John Wiley and Sons, New York.

Pietsch, T.W. 1995. *Historical portrait of the progress of ichthyology from its origin to our own time, by Georges Cuvier.* Johns Hopkins University Press, Baltimore.

Platnick, N.I. 1976. Concepts of dispersal in historical biogeography. *Systematic Zoology,* 25, 294–295.

Platnick, N.I. 1979. Philosophy and the transformation of cladistics. *Systematic Zoology,* 28, 537–546.

Platnick, N.I. 1991. On areas of endemism. *Australian Systematic Botany,* 4, xi–xii.

Platnick, N. I., and G. Nelson. 1978. A method of analysis for historical biogeography. *Systematic Zoology*, 27, 1–16.

Platts, J. 2006. Newton, Goethe and the process of perception: An approach to design. *Optics and Laser Technology*, 38, 205–209.

Plovanich, A. E., and J. L. Panero. 2004. A phylogeny of the ITS and ETS for *Montanoa* (Asteraceae: Heliantheae). *Molecular Phylogenetics and Evolution*, 31, 815–821.

Porzecanski, A. L., and J. Cracraft. 2005. Cladistic analysis of distributions and endemism (CADE): Using raw distributions of birds to unravel the biogeography of the South American aridlands. *Journal of Biogeography*, 32, 261–275.

Possingham, H. P., and K. A. Wilson. 2005. Turning up the heat on hotspots. *Nature*, 436, 919–920.

Powell, A. W. B. 1957. Marine provinces of the Indo-West Pacific. In *Proceedings of the Eighth Pacific Science Congress of the Pacific Science Association, volume III* (pp. 359–370). *Oceanography*. National Research Council of the Philippines, Quezon City.

Prance, G. T. 1982. *Biological diversification in the tropics*. Columbia University Press, New York.

Price, D. J. D S. 1963. *Little science, big science*. Columbia University Press, New York.

Prichard, J. C. 1826. *Researches into the physical history of mankind, second edition*. Houlfton and Stoneman, London.

Quijano-Abril, M., R. Callejas-Posada, and D. R. Miranda-Esquivel. 2006. Areas of endemism and distribution patterns for neotropical Piper species (Piperaceae). *Journal of Biogeography*, 33, 1266–1278.

Ragan, M. A. 1998. On the delineation and higher-level classification of algae. *European Journal of Phycology*, 33, 1–15.

Rahbek, C., N. J. Gotelli, R. K. Colwell, G. L. Entsminger, T. F. L. V. B. Rangel, and G. R. Graves. 2007. Predicting continental-scale patterns of bird species richness with spatially explicit models. *Proceedings of the Royal Society of London B*, 274, 165–174.

Ratzel, F. 1891. *Anthropogeographie, volume* 2. Engelhorn, Stuttgart.

Renema, W., D. R. Bellwood, J. C. Braga, K. Bromfield, R. Hall, K. G. Johnson, P. Lunt, C. P. Meyer, L. B. McMonagle, R. J. Morley, A. O'Dea, J. A. Todd, F. P. Wesselingh, M. E. J. Wilson, and J. M. Pandolfi. 2008. Hopping hotspots: Global shifts in marine biodiversity. *Science*, 321, 654–657.

Ricklefs, R. E. 2004. A comprehensive framework for global patterns in biodiversity. *Ecology Letters*, 7, 1–15.

Riddle, B. R. 2005. Is biogeography emerging from its identity crisis? *Journal of Biogeography*, 32, 185–186.

Riddle, B. R., and D. J. Hafner. 2004. The past and future roles of phylogeography in historical biogeography. In M. V. Lomolino and L. R. Heaney (eds.), *Frontiers of biogeography* (pp. 93–110). Sinauer Associates, Sunderland, Massachusetts.

Ronquist, F. 1997. Dispersal-vicariance analysis: A new approach to quantification of historical biogeography. *Systematic Biology*, 46, 195–203.

Ronquist, F. 1998. Three-dimensional cost-matrix optimization and maximum co-speciation. *Cladistics,* 14, 167–172.

Rosen, B.R. 1985. Long-term geographical controls on regional diversity. *The Open University Geological Society Journal,* 6, 25–30.

Rosen, B.R. 1988. From fossils to Earth history: Applied historical biogeography. In A.A. Myers and P.A. Giller (eds.), *Analytical biogeography: An integrated approach to the study of animal and plant distributions* (pp. 437–481). Chapman and Hall, London.

Rosen, B.R., and A.B. Smith. 1988. Tectonics from fossils? Analysis of reef-coral and sea-urchin distributions form Late Cretaceous to Recent, using a new method. In M.G. Audley-Charles and A. Hallam (eds.), *Gondwana and Tethys* (pp. 275–306). Geological Society of London Special Publication.

Rosen, D.E. 1974. The phylogeny and zoogeography of salmoniform fishes and the relationships of *Lepidogalaxias salamandroides. Bulletin of the American Museum of Natural History,* 153, 265–326.

Rosen, D.E. 1978. Vicariant patterns and historical explanation in biogeography. *Systematic Zoology,* 27, 159–188.

Rosen, D.E. 1979. Fishes from the uplands and intermontane basins of Guatemala: Revisionary studies and comparative geography. *Bulletin of the American Museum of Natural History,* 162, 267–376.

Roughgarden, J. 1995. *Anolis lizards of the Caribbean: Ecology, evolution and plate tectonics.* Oxford University Press, New York.

Rudwick, M.J.S. 2005. *Bursting the limits of time: The reconstruction of geohistory in the age of revolution.* University of Chicago Press, Chicago.

Sanderson, M.J. 2005. Where have all the clades gone? A systematist's take on *Inferring Phylogenies. Evolution,* 59, 2056–2058.

Santini, F., and R. Winterbottom. 2002. Historical biogeography of Indo-western Pacific coral reef biota: Is the Indonesian region a centre of origin? *Journal of Biogeography,* 29, 189–205.

Santos, C.M.D., and D.S. Amorim. 2007. Why biogeographical hypotheses need a well supported phylogenetic framework: A conceptual evaluation. *Papéis Avulsos de Zoologia,* 47, 63–73.

Sax, D., B.P. Kinlan, and K.F. Smith. 2005. A conceptual framework for comparing species assemblages in native and exotic habitats. *Oikos,* 108, 457–464.

Scharff, R.F. 1911. *Distribution and origin of life in America.* Constable and Company, London.

Schilder, F.A. 1956. *Lehrbuch der allegemeinen zoogeographie.* VEB Gustav Fischer Verlag, Jena.

Schilder, F.A., and M. Schilder. 1938. Prodome of a monograph on living Cypraeidae. *Proceedings of the Malacaological Society of London,* 23, 119–180.

Schilder, F.A., and M. Schilder. 1939. Prodome of a monograph on living Cypraeidae. *Proceedings of the Malacaological Society of London,* 23, 181–252.

Schmarda, L.K. 1853. *Die geographische verbreitung der thiere.* Carl Gerold and Son, Vienna.

Schönswetter, P., A. Tribsch, M. Barfuss, and H. Niklfeld. 2002. Several Pleistocene refugia detected in the high alpine plant *Phyteuma globulariifolium*

Sternb. & Hoppe (Campanulaceae) in the European Alps. *Molecular Ecology,* 11, 2637–2647.

Sclater, P.L. 1858. On the general geographical distribution of the members of the class Aves. *Journal of the Proceedings of the Linnean Society: Zoology,* 2, 130–145.

Sclater, P.L. 1864. The mammals of Madagascar. *The Quarterly Journal of Science,* 1, 213–219.

Seberg, O. 1986. A critique of the theory and methods of panbiogeography. *Systematic Zoology,* 35, 369–380.

Seberg, O. 1991. Biogeographic congruence in the South Pacific. *Australian Systematic Botany,* 4, 127–136.

Shields, O. 1979. Evidence for the initial opening of the Pacific Ocean in the Jurassic. *Palaeogeography, Palaeoclimatography, Palaeoecology,* 26, 181–220.

Shields, O. 1983. Trans-Pacific biotic links that suggest Earth expansion. In S.W. Carey (ed.), *Expanding Earth symposium* (pp. 199–205). University of Tasmania, Hobart.

Shields, O. 1991. Pacific biogeography and rapid Earth expansion. *Journal of Biogeography,* 18, 583–585.

Shields, O. 1996. Plate tectonics or an expanding Earth? *Journal of the Geological Society of India,* 47, 399–408.

Shubin, N. 2008. *Your inner fish: A journey into the 3.5-billion-year history of the human body.* Pantheon Books, New York.

Siddall, M.E. 2004. Fallacies of false attribution: The defense of BPA by Brooks, Dowling, van Veller, and Hoberg. *Cladistics,* 20, 376–377.

Siddall, M.E., and S.L. Perkins. 2003. Brooks parsimony analysis: A valiant failure. *Cladistics,* 19, 554–564.

Siebert, D.J. 1992. Tree statistics; trees and 'confidence'; consensus trees; alternatives to parsimony; character weighting; character conflict and its resolution. In P.L. Forey, C.J. Humphries, I.L. Kitching, R.W. Scotland, D.J. Siebert, and D.M. Williams (eds.), *Cladistics: A practical course in systematics* (pp. 72–88). The Systematics Association Publication No. 10, Oxford University Press, Inc., New York.

Simpson, G.G. 1940. Mammals and land bridges. *Journal of the Washington Academy of Science,* 30, 137–163.

Simpson, G.G. 1950. History of the fauna of Latin America. *American Science,* 38, 361–389.

Smith, C.H. 1989. Historical biogeography: geography as evolution, evolution as geography. *New Zealand Journal of Zoology,* 16, 773–785.

Smith, A.G., and A. Hallam. 1970. The fit of the southern continents. *Nature,* 225, 139–144.

Sokal, R.R. 1979. Testing statistical significance of geographic variation patterns. *Systematic Zoology,* 28, 227–232.

Spalding, M.D., H.E. Fox, G.R. Allen, N. Davidson, Z.A. Ferdaña, M. Finlayson, B.S. Halpern, M.A. Jorge, A. Lombana, S.A. Lourie, K.D. Martin, E. McManus, J. Molnar, C.A. Recchia, and J. Robertson. 2007. Marine ecoregions of the world: A bioregionalization of coastal and shelf areas. *Bioscience,* 57, 573–583.

Sparks, J.S., and W.L. Smith. 2004. Phylogeny and biogeography of cichlid fishes. *Cladistics,* 20, 501–517.

Spellerberg, I.F., and J.W.D. Sawyer. 1999. *An introduction to applied biogeography.* Cambridge University Press, Cambridge.

Springer, V.G. 1982. Pacific plate biogeography, with special reference to shorefishes. *Smithsonian Contributions to Zoology,* 367, 1–182.

Stehli, F.G., and J.W. Wells. 1971. Diversity and age patterns in hermatypic corals. *Systematic Zoology,* 20, 115–126.

Stejneger, L. 1901. Scharff's history of the European fauna. *The American Naturalist,* 35, 87–116.

Stiassny, M.L.J. 1981. The phyletic status of the family Cichlidae (Pisces, Perciformes): A comparative anatomical investigation. *Netherlands Journal of Zoology,* 31, 275–314.

Stiassny, M.L.J. 1991. Phylogenetic intrarelationships of the family Cichlidae: An overview. In M.H.A. Keenleyside (ed.), *Cichlid fishes: Behaviour, ecology and evolution* (pp. 1–35). Chapman and Hall, London.

Stuessy, T. 2000. Taxon names are not defined. *Taxon,* 49, 231–233.

Suess, E. 1885. *Das antlitz der Erde, volume* 3. Braunmüller, Wien.

Swofford, D.L. 1991. When are phylogeny estimates from molecular and morphological data incongruent? In M.M. Miyamoto and J. Cracraft (eds.), *Phylogenetic analysis of DNA sequences* (pp. 295–333). Oxford University Press, New York.

Szostak, R. 2005. *Classifying science: Phenomena, data, theory, method, practice.* Information Science and Knowledge Management, Springer, New York.

Szumik, C.A., and P.A. Goloboff. 2004. Areas of endemism: An improved optimality criterion. *Systematic Biology,* 53, 968–977.

Takhtajan, A. 1986. *Floristic regions of the world.* University of California Press, Berkeley.

Tarling, D.H. 1972. Another Gondwanaland. *Nature,* 238, 92–93.

Tarling, D.[H.], and M. Tarling. 1975. *Continental drift: A study of the Earth's moving surface.* Anchor Books, New York.

Taylor, D.J., T.L. Finston, and P.D.N. Hebert. 1998. Biogeography of a widespread freshwater crustacean: Pseudocongruence and cryptic endemism in the North American *Daphnia laevis* complex. *Evolution,* 52, 1648–1670.

Templeton, A.R. 1998. Nested clade analysis of phylogeographic data: Testing hypothesis about gene flow and population history. *Molecular Ecology,* 7, 381–397.

Teske, P.R., M.I. Cherry, and C.A. Matthee. 2004. The evolutionary history of seahorses (Syngnathidae: Hippocampus): Molecular data suggest a West Pacific origin and two invasions of the Atlantic Ocean. *Molecular Phylogenetics and Evolution,* 30, 273–286.

Thompson, D.W. 1910. *The works of Aristotle translated into English.* Oxford University Press, Oxford.

Trewick, S.A., and R.H. Cowie (eds.). 2008. Evolution on Pacific islands: Darwin's legacy. *Philosophical Transactions of the Royal Society B, Biological Sciences,* 363(1508), 3287–3465.

Udvardy, M.D.F. 1975. A classification of the biogeographical provinces of the world. IUCN Occasional Papers. International Union for Conservation of Nature and Natural Resources, Morges, Switzerland.

Van Balgooy, M.M.J. 1971. Plant-geography of the Pacific, as based on a census of Phanerogam genera. *Blumea* Supplement, 6, 1–222.

Vane-Wright, R.I. 1991. Transcending the Wallace Line: Do the western edges of the Australian region and the Australian Plate coincide? *Australian Systematic Botany*, 4, 183–197.

Vences, M., J. Freyhof, R. Sonnenberg, J. Kosuch, and M. Veith. 2001. Reconciling fossils and molecules: Cenozoic divergence of cichlid fishes and the biogeography of Madagascar. *Journal of Biogeography*, 28, 1091–1099.

Vermeij, G. 2002. The geography of evolutionary opportunity: Hypothesis and two cases in gastropods. *Integrative and Comparative Biology*, 42, 935–940.

Viloria, Á.L. 2004. Biogeografía, la dimensión espacial de la Evolución. *Intercienca*, 29, 163–164.

Viloria, Á.L. 2005. Las mariposas (Lepidoptera: Papilionoidea) y la regionalización biogeográfica de Venezuela. In J.E. Llorente and J.J. Morrone (eds.), *Regionalización biogeográfica en Iberoamérica y tópicos afines* (pp. 441–459). Primeras Jornadas Biogeográficas de la Red Iberoamericana de Biogeografía y Entomología Sistemática (RIBES XII.I-CYTED), Las Prensas de Ciencias, Facultad de Ciencias, UNAM, México, DF.

Wagner, W.L., and V.A. Funk (eds.). 1995. Hawaiian biogeography: Evolution on a hot spot archipelago. Smithsonian Institution Press, Washington, D.C.

Wagstaff, S.J., and M.I. Dawson. 2000. Classification, origin, and patterns of diversification of *Corynocarpus* (Corynocarpaceae) inferred from DNA sequences. *Systematic Botany*, 25, 134–149.

Wallace, A.R. 1869. *The Malay Archipelago: The land of the orang-utan and the bird of paradise. A narrative of travel, with sketches of man and nature.* Dover Publications, New York [1962 reprint].

Wallace, A.R. 1876. *The geographical distribution of animals; with a study of the relations of living and extinct faunas as elucidating the past changes of the Earth's surface.* Macmillan, London.

Wallace, A.R. 1881. *Island life, or the phenomenon and causes of insular faunas and floras including a revision and attempted solution of the problem of geological climates.* Prometheus Books, New York [1998 reprint].

Wallace, A.R. 1894. What are zoological regions? *Nature*, 49, 610–613.

Wallace, C.C. 1997. The Indo-Pacific centre of coral diversity re-examined at species level. In H.A. Lessios and I.G. Macintyre (eds.), *Proceedings of the Eighth International Coral Reef Symposium* (pp. 365–370). Smithsonian Tropical Research Institute, Panama.

Wallace, C.C., J.M. Pandolfi, A. Young, and J. Wolstenholme. 1991. Indo-Pacific coral biogeography: A case study from the *Acropora selago* group. *Australian Systematic Botany*, 4, 199–210.

Wallace, C. C. and B.R. Rosen. 2006. Diverse staghorn corals (Acropora) in high-latitude Eocene assemblages: implications for the evolution of modern diversity patterns of reef corals. *Proceedings of the Royal Society B, Biological Sciences*, 273(1589), 975–982.

Wegener, A. 1915. *Die enstehung der kontinente und ozeane.* F. Vieweg and Sohn, Braunschweig.

Wegener, A. 1929. *Die enstehung der kontinente und ozeane, fourth edition.* F. Vieweg and Sohn, Braunschweig.

Weinberg, A.M. 1961. Impact of large-scale science on the United States. *Science,* 134, 164.

Westermann, G.E.G. 2000. Biochore classification and nomenclature in paleobiogeography: An attempt at order. *Palaeogeography, Palaeoclimatography, Palaeoecology,* 158, 1–13.

Whittaker, R.J., K.A. Triantis, and R.J. Ladle. 2008. A general dynamic theory of oceanic island biogeography. *Journal of Biogeography,* 35, 977–994.

Wiley, E. 1979. Cladograms and phylogenetic trees. *Systematic Zoology,* 28, 88–92.

Wiley, E.O. 1987. Methods in vicariance biogeography. In P. Hovencamp (ed.), *Systematics and evolution: A matter of diversity* (pp. 283–306). University of Utrecht, Utrecht.

Wilkins, J.S. 2008. *Defining species: A sourcebook from antiquity to today.* Peter Lang Press, Bern.

Williams, D.M. 2006. Ernst Haeckel and Louis Agassiz: Trees that bite and their geographical dimension. In M.C. Ebach and R.S. Tangney (eds.), *Biogeography in a changing world* (pp. 1–59). Fifth Biennial Conference of the Systematics Association, UK, CRC Press, Boca Raton, Florida.

Williams, D.M., and M.C. Ebach. 2008. *Foundations of systematics and biogeography.* Springer, New York.

Williams, S.T., and D.G. Reid. 2004. Speciation and diversity on tropical rocky shores: A global phylogeny of snails of the genus *Echinolittorina. Evolution,* 58, 2227–2251.

Wojcicki, M., and D.R. Brooks. 2005. PACT: An efficient and powerful algorithm for generating area cladograms. *Journal of Biogeography,* 32, 755–774.

Wolfram, S. 2002. *A new kind of science.* Wolfram Media, Inc., Champaign, Illinois.

Wolfson, A. 1948. Bird migration and the concept of continental drift. *Science,* 108, 23–30.

Young, G.C. 1984. Comments on the phylogeny and biogeography of antiarchs (Devonian placoderm fishes), and the use of fossils in biogeography. *Proceedings of the Linnean Society of New South Wales,* 107, 443–473.

Young, G.C. 1995. Application of cladistics to terrane history—parsimony analysis of qualitative geological data. *Journal of Southeast Asian Earth Sciences,* 11, 167–176.

Zandee, M., and M. Roos. 1987. Component compatibility in historical biogeography. *Cladistics,* 3, 305–332.

Zaragüeta Bagils, R., H. Lelièvre, and P. Tassy. 2004. Temporal paralogy, cladograms, and the quality of the fossil record. *Geodiversitas,* 26, 381–389.

Zimmermann, E.A.W.V. 1777. *Specimen zoologiae geographicae, quadrupedum domocilia et migrationes sistems.* Theodorum Haak, Leiden/Batavia.

Zuckerkandl, E., and L. Pauling. 1962. Molecular disease, evolution, and genic heterogeneity. In M. Kasha and B. Pullman (eds.), *Horizons in biochemistry* (pp. 189–225) Academic Press, New York.

Index

Note: A page number followed by b, f, m, n, *or* t, *indicates a box, figure, map, note, or table, respectively.*

About the Authors

Lynne Parenti is Curator of Fishes and Research Scientist at the Smithsonian's National Museum of Natural History. Lynne developed an early fascination with natural history, especially comparative vertebrate anatomy and biogeography, while exploring natural and altered habitats of her native New York City. Fieldwork has since taken her to many and varied tropical and temperate sites, including New Guinea Sulawesi, Borneo, Tasmania, China, Taiwan, Surinam and Cuba. Lynne was the first woman ichthyologist elected President of the American Society of Ichthyologists and Herpetologists. As adjunct professor at the George Washington University, she directs the research of graduate students in systematic ichthyology.

Malte Ebach is a Postdoctoral Research Fellow at Arizona State University. Since childhood, Malte has been interested in natural history, especially birds and fossils, and early on became an avid collector of natural history objects ranging from insect specimens to historical science books. Wanting to roll all these interests into one, he embraced the field of biogeography while majoring in Geology and Geography at the University of Newcastle, Australia. Malte has published extensively on biogeography, its methods, theory, and history, and currently serves on the editorial board of the *Journal of Biogeography*.